I0500476

Overview

"The Frequency Factor: Unraveling the Impact on Sleep, Weight Loss, and Well-Being" is a comprehensive guide that explores the profound influence of frequency on various aspects of our lives. In this book, readers will gain a deep understanding of how frequency affects sleep patterns, weight loss efforts, stress levels, cognitive function, mood, physical performance, immune system function, and overall well-being. Through a scientific lens, the book delves into the intricate relationship between frequency and sleep quality, metabolism, stress management, cognitive function, and physical health. It provides evidence-based strategies and techniques for optimizing sleep patterns, enhancing weight loss efforts, managing stress, promoting relaxation, and boosting overall well-being through frequency-based practices. With practical tips and insights, readers will learn how to incorporate frequency into their daily lives, create a frequency-based routine, cultivate mindfulness, and sustain these practices for long-term well-being. Whether you are seeking better sleep, effective weight loss strategies, stress reduction techniques, improved mental health, enhanced physical performance, or an overall sense of well-being, "The Frequency Factor" offers valuable knowledge and tools to unlock the transformative power of frequency in your life.

Table Of Contents

1 Understanding the Power of Frequency

1.1 Introduction to Frequency and its Effects

Frequency is a fundamental concept in the study of physics and is defined as the number of occurrences of a repeating event per unit of time. While frequency is commonly associated with sound and electromagnetic waves, it also plays a crucial role in various aspects of our lives, including sleep, weight loss, and overall well-being. Understanding the impact of frequency on these areas can provide valuable insights into optimizing our health and enhancing our quality of life.

In this chapter, we will delve into the fascinating world of frequency and explore its effects on sleep, weight loss, and well-being. We will uncover the scientific principles behind frequency and its influence on different physiological processes. By gaining a deeper understanding of how frequency affects our bodies, we can harness its power to improve our sleep patterns, support weight loss efforts, manage stress, enhance cognitive function, and promote overall well-being.

1.1.1 The Science Behind Frequency and Sleep

Sleep is a vital process that allows our bodies to rest, repair, and rejuvenate. The quality and duration of our sleep have a profound impact on our physical and mental health. Research has shown that frequency plays a significant role in regulating our sleep patterns.

Our brain operates on different frequency bands, known as brainwaves, which are associated with different states of consciousness. These brainwaves include delta waves (0.5-4 Hz), theta waves (4-8 Hz), alpha waves (8-12 Hz), beta waves (12-30 Hz), and gamma waves (30-100 Hz). Each of these brainwave frequencies corresponds to a specific mental state, such as deep sleep, dreaming, relaxation, focus, and heightened awareness.

During sleep, our brain transitions through these different brainwave frequencies in a cyclical pattern. The frequency and duration of each brainwave stage determine the overall quality of our sleep. Disruptions in these patterns can lead to sleep disorders, such as insomnia or sleep apnea, and have a negative impact on our well-being.

1.1.2 Frequency's Impact on Sleep Quality

The frequency of brainwave activity during sleep directly influences the quality of our rest. For example, the presence of delta waves is associated with deep sleep, which is crucial for physical restoration and immune system function. On the other hand, an excess of beta waves, which are associated with wakefulness and alertness, can disrupt the transition into deep sleep and result in a restless night.

Moreover, external factors, such as exposure to certain frequencies of light and sound, can also affect our sleep quality. For instance, exposure to blue light, which has a higher frequency, emitted by electronic devices before bedtime can suppress the production of melatonin, a hormone that regulates sleep-wake cycles. This can lead to difficulty falling asleep and disrupted sleep patterns.

1.1.3 Optimizing Sleep Patterns with Frequency

Understanding the relationship between frequency and sleep can help us optimize our sleep patterns and improve the quality of our rest. By utilizing techniques such as binaural beats, which involve listening to two slightly different frequencies in each ear, we can entrain our brainwaves to desired frequencies and promote relaxation and sleep.

For example, listening to binaural beats in the delta frequency range can help induce deep sleep and enhance the restorative aspects of our sleep. Similarly, using binaural beats in the theta frequency range can facilitate relaxation and promote a calm state of mind before sleep.

In addition to binaural beats, other frequency-based techniques, such as isochronic tones and white noise, can also be used to optimize sleep patterns. These techniques work by providing a consistent frequency or sound that helps mask external disturbances and promote a more peaceful sleep environment.

By incorporating frequency-based practices into our bedtime routine, we can create a conducive environment for quality sleep and reap the benefits of improved rest and overall well-being.

In the next section, we will explore the relationship between frequency and weight loss, uncovering how frequency affects metabolism and how it can be used to enhance weight loss efforts.

1.2 The Science Behind Frequency and Sleep

Sleep is a fundamental aspect of our daily lives, and its quality and duration have a significant impact on our overall well-being. The science behind frequency and sleep reveals a fascinating connection between the two. In this section, we will explore how frequency affects sleep patterns, sleep quality, and ultimately, our overall health.

The Role of Brainwaves in Sleep

To understand the science behind frequency and sleep, we must first delve into the concept of brainwaves. Brainwaves are electrical impulses generated by the brain, and they can be categorized into different frequencies, each associated with a specific state of consciousness. The four primary types of brainwaves are delta, theta, alpha, and beta.

Delta waves are the slowest brainwaves, with a frequency range of 0.5 to 4 Hz. They are typically observed during deep sleep and are associated with restorative processes in the body, such as tissue repair and hormone regulation.

Theta waves have a frequency range of 4 to 8 Hz and are present during light sleep, as well as during the early stages of deep sleep. They are associated with dreaming, creativity, and emotional processing.

Alpha waves have a frequency range of 8 to 13 Hz and are present when we are in a relaxed and wakeful state. They are often associated with a calm and focused mind.

Beta waves have the highest frequency range, typically above 13 Hz, and are present when we are awake and engaged in active mental or physical tasks. They are associated with alertness, concentration, and problem-solving.

Frequency and Sleep Patterns

The frequency of brainwaves plays a crucial role in determining our sleep patterns. When we are awake, our brain predominantly produces beta waves. As we transition into sleep, the frequency of our brainwaves slows down, with alpha waves giving way to theta waves and eventually delta waves during deep sleep.

Research has shown that exposure to specific frequencies can influence our brainwave patterns and, consequently, our sleep. For example, studies have demonstrated that listening to music or sounds with a slow tempo and low-frequency range can help induce relaxation and promote the transition into sleep. This is because these low-frequency sounds can entrain our brainwaves to match their rhythm, facilitating a shift from beta to alpha and theta waves.

Conversely, exposure to high-frequency sounds or stimuli can have an alerting effect on the brain, making it more difficult to fall asleep or maintain deep sleep. This is why it is recommended to create a sleep environment that is free from disruptive noises, such as loud alarms or electronic devices emitting high-frequency sounds.

Frequency and Sleep Quality

The impact of frequency on sleep quality extends beyond brainwave patterns. It also influences the release of hormones and neurotransmitters that regulate sleep and wakefulness. For example, exposure to natural light, which contains a broad spectrum of frequencies, helps regulate our circadian rhythm, the internal clock that governs our sleep-wake cycle.

Natural light, particularly in the blue spectrum, suppresses the production of melatonin, a hormone that promotes sleep. This is why exposure to bright light, especially in the morning, can help reset our internal clock and improve sleep quality. On the other hand, exposure to artificial light, particularly from electronic devices emitting blue light, can disrupt our circadian rhythm and interfere with the onset of sleep.

Furthermore, studies have shown that exposure to specific frequencies, such as those emitted by white noise machines or pink noise, can improve sleep quality by masking disruptive sounds and promoting a more relaxed state of mind. These frequencies create a soothing background noise that can drown out external disturbances and help individuals fall asleep faster and experience deeper, more restorative sleep.

The Importance of Frequency Synchronization

Another fascinating aspect of the science behind frequency and sleep is the concept of frequency synchronization. When we are in a state of deep sleep, our brainwaves tend to synchronize with other physiological processes in the body, such as heart rate, breathing, and even the release of growth hormones.

Research suggests that certain frequencies, such as those found in binaural beats or isochronic tones, can help induce frequency synchronization in the brain. Binaural beats are created by playing two slightly different frequencies in each ear, while isochronic tones involve the use of evenly spaced pulses of sound or light.

By listening to binaural beats or isochronic tones that correspond to the desired brainwave frequency, individuals can potentially enhance their sleep experience. For example, listening to binaural beats in the delta range may help promote deep sleep and facilitate the release of growth hormones, which are essential for tissue repair and regeneration.

In conclusion, the science behind frequency and sleep reveals a fascinating interplay between brainwaves, sleep patterns, and overall well-being. By understanding how different frequencies affect our sleep, we can optimize our sleep environment, adopt healthy sleep hygiene practices, and explore frequency-based techniques to enhance the quality and duration of our sleep. In the following sections, we will further explore the impact of frequency on sleep quality, weight loss, stress management, and overall well-being.

1.3 Frequency's Impact on Sleep Quality

Sleep is a fundamental aspect of our overall well-being, and the quality of our sleep directly affects our physical and mental health. The impact of frequency on sleep quality is a fascinating area of study that has gained significant attention in recent years. In this section, we will explore how frequency influences our sleep patterns and discuss the various ways in which it can either enhance or disrupt our sleep quality.

The Role of Frequency in Sleep Regulation

To understand the impact of frequency on sleep quality, it is essential to first grasp the concept of sleep regulation. Sleep regulation refers to the complex processes that govern our sleep-wake cycle, ensuring that we experience the right amount of sleep at the appropriate times. These processes are influenced by various factors, including external cues, internal body signals, and environmental conditions.

Frequency plays a crucial role in sleep regulation by influencing the synchronization of our biological rhythms. Our bodies have internal clocks, known as circadian rhythms, which regulate our sleep-wake cycle. These rhythms are influenced by external cues, such as light and darkness, and help us maintain a consistent sleep pattern.

Frequency's Influence on Sleep Patterns

The frequency of external stimuli, such as light and sound, can significantly impact our sleep patterns. Exposure to certain frequencies of light, particularly blue light emitted by electronic devices, can disrupt our natural circadian rhythms and suppress the production of melatonin, a hormone that promotes sleep. This disruption can lead to difficulties falling asleep, staying asleep, and achieving restorative sleep.

Similarly, exposure to loud or disruptive sounds at certain frequencies can also disturb our sleep. Noise pollution, such as traffic noise or loud neighbors, can trigger physiological responses that disrupt our sleep cycles and reduce sleep quality. These disruptions can result in fragmented sleep, decreased sleep efficiency, and increased awakenings throughout the night.

Frequency's Effect on Sleep Architecture

Sleep architecture refers to the structure and organization of our sleep cycles, including the different stages of sleep we experience throughout the night. The impact of frequency on sleep architecture is significant, as it can influence the duration and distribution of each sleep stage.

Research has shown that exposure to specific frequencies, such as low-frequency sounds or vibrations, can promote deep sleep and enhance sleep quality. Deep sleep is a crucial stage of sleep associated with physical restoration, memory consolidation, and overall well-being. By optimizing the frequency of external stimuli, we can potentially enhance the duration and quality of deep sleep, leading to improved overall sleep quality.

Conversely, exposure to high-frequency sounds or disturbances can disrupt our sleep architecture and reduce the amount of time spent in restorative sleep stages. This disruption can result in increased sleep fragmentation, decreased sleep efficiency, and a higher likelihood of experiencing sleep disorders such as insomnia or sleep apnea.

Frequency-Based Strategies for Improving Sleep Quality

Understanding the impact of frequency on sleep quality opens up opportunities for utilizing frequency-based strategies to optimize our sleep patterns. Here are some techniques that can help improve sleep quality through frequency:

1. **White Noise and Pink Noise**: White noise, which contains equal energy across all frequencies, and pink noise, which emphasizes lower frequencies, can help mask disruptive sounds and promote a more peaceful sleep environment.
2. **Binaural Beats**: Binaural beats are created by playing two slightly different frequencies in each ear, resulting in the perception of a third frequency. These beats have been found to promote relaxation and improve sleep quality when used during bedtime.
3. **Light Therapy**: Exposure to specific frequencies of light, such as bright light in the morning and warm, dim light in the evening, can help regulate our circadian rhythms and promote a healthy sleep-wake cycle.
4. **Mindfulness and Meditation**: Practicing mindfulness and meditation techniques that focus on breath awareness or body scanning can help calm the mind and prepare the body for sleep.
5. **Creating a Sleep-Friendly Environment**: Minimizing exposure to disruptive frequencies, such as turning off electronic devices or using earplugs, can create a more conducive sleep environment.

By incorporating these frequency-based strategies into our daily routines, we can optimize our sleep patterns and enhance the quality of our sleep, leading to improved overall well-being.

In the next section, we will delve deeper into the topic of optimizing sleep patterns with frequency, exploring additional techniques and practices that can further enhance our sleep quality and promote restorative sleep.

1.4 Optimizing Sleep Patterns with Frequency

Sleep is a vital component of our overall well-being, and the quality and duration of our sleep can have a significant impact on our physical and mental health. In recent years, researchers have begun to explore the role of frequency in optimizing sleep patterns and improving sleep quality. This section will delve into the various ways in which frequency can be utilized to enhance our sleep and promote better overall well-being.

1.4.1 Understanding the Role of Frequency in Sleep

Before we can explore how frequency can optimize sleep patterns, it is important to understand the role of frequency in our sleep-wake cycle. Our bodies naturally follow a circadian rhythm, which is a 24-hour internal clock that regulates our sleep and wakefulness. This rhythm is influenced by various factors, including exposure to light, physical activity, and even the foods we consume.

Frequency, in the context of sleep, refers to the specific wavelengths of sound, light, or electromagnetic waves that can impact our brain activity and subsequently affect our sleep patterns. Different frequencies have different effects on our brainwaves, and understanding these effects can help us optimize our sleep.

1.4.2 The Impact of Frequency on Sleep Quality

Research has shown that certain frequencies can have a profound impact on the quality of our sleep. For example, studies have found that exposure to low-frequency sounds, such as those produced by white noise machines or calming music, can help promote relaxation and improve sleep quality. These low-frequency sounds have been found to slow down brainwave activity, leading to a more restful and rejuvenating sleep experience.

On the other hand, high-frequency sounds, such as those produced by loud noises or electronic devices, can disrupt our sleep and lead to poorer sleep quality. These high-frequency sounds can stimulate brainwave activity, making it more difficult to fall asleep and stay asleep throughout the night.

1.4.3 Using Frequency to Optimize Sleep Patterns

Now that we understand the impact of frequency on sleep quality, let's explore how we can utilize frequency to optimize our sleep patterns. One effective technique is to create a sleep environment that is conducive to relaxation and tranquility. This can be achieved by incorporating low-frequency sounds, such as gentle nature sounds or calming music, into our bedtime routine. These soothing sounds can help to drown out any disruptive noises and create a peaceful atmosphere that promotes better sleep.

In addition to sound frequencies, light frequencies also play a crucial role in our sleep patterns. Exposure to bright, blue light emitted by electronic devices, such as smartphones and tablets, can interfere with our natural sleep-wake cycle and make it harder to fall asleep. To optimize sleep patterns, it is recommended to limit exposure to electronic devices before bedtime and instead opt for dimmer, warmer lighting in the evening.

1.4.4 Frequency-Based Techniques for Better Sleep

In addition to creating a sleep-friendly environment, there are several frequency-based techniques that can be incorporated into our bedtime routine to optimize sleep patterns. One such technique is binaural beats, which involve listening to two slightly different frequencies in each ear. This creates a third frequency in the brain, known as the binaural beat, which can help induce a state of relaxation and promote better sleep.

Another technique is the use of brainwave entrainment, which involves listening to specific frequencies that align with different stages of sleep. By listening to these frequencies, our brainwaves can synchronize with the desired frequency, leading to a more restful and rejuvenating sleep experience.

Furthermore, mindfulness meditation, which involves focusing on the present moment and cultivating a sense of calm and relaxation, can also be beneficial for optimizing sleep patterns. By practicing mindfulness before bed, we can quiet our minds and prepare ourselves for a restful night's sleep.

Conclusion

Optimizing sleep patterns is essential for our overall well-being, and frequency can play a significant role in achieving this goal. By understanding the impact of frequency on sleep quality and incorporating frequency-based techniques into our bedtime routine, we can enhance our sleep and wake up feeling refreshed and rejuvenated. In the next chapter, we will explore the relationship between frequency and weight loss, and how frequency can be utilized to support our weight management efforts.

2 Frequency and Weight Loss

2.1 Exploring the Relationship Between Frequency and Weight

Frequency, in the context of this book, refers to the vibrations or oscillations that occur in various aspects of our lives, including our physical and mental well-being. In this chapter, we will delve into the relationship between frequency and weight, and how understanding this connection can help us optimize our weight loss efforts.

Weight loss is a complex process influenced by various factors such as diet, exercise, genetics, and metabolism. However, recent research has shed light on the role of frequency in weight management. It appears that the frequency at which our body functions can have a significant impact on our ability to lose weight and maintain a healthy body composition.

2.1.1 The Influence of Frequency on Metabolism

Metabolism is the process by which our body converts food into energy. It plays a crucial role in weight management, as a higher metabolic rate allows us to burn more calories throughout the day. Interestingly, studies have shown that frequency can influence our metabolism.

One aspect of frequency that affects metabolism is the frequency of our meals. Research suggests that eating smaller, more frequent meals throughout the day can boost our metabolism compared to consuming fewer, larger meals. This is because frequent meals help to stabilize blood sugar levels and prevent spikes and crashes, which can negatively impact metabolism.

Additionally, the frequency at which we engage in physical activity also plays a role in our metabolism. Regular exercise, especially high-intensity interval training (HIIT), has been shown to increase metabolic rate both during and after exercise. By incorporating regular exercise sessions into our routine, we can elevate our metabolic rate and enhance our weight loss efforts.

2.1.2 Harnessing Frequency for Weight Loss

Understanding the relationship between frequency and weight loss allows us to leverage this knowledge to optimize our weight management strategies. Here are some ways in which we can use frequency to enhance our weight loss efforts:

1. Meal Frequency and Timing

As mentioned earlier, consuming smaller, more frequent meals can help boost metabolism. Additionally, paying attention to the timing of our meals can also have an impact. Research suggests that having a consistent meal schedule, with meals spaced evenly throughout the day, can help regulate our metabolism and support weight loss.

2. Mindful Eating

Practicing mindful eating involves paying attention to the frequency and quality of our food intake. By being present and fully engaged in the act of eating, we can better listen to our body's hunger and fullness cues. This can prevent overeating and promote a healthier relationship with food, ultimately supporting weight loss goals.

3. Regular Exercise Routine

Incorporating regular exercise into our routine is crucial for weight loss. By engaging in physical activity at a frequency that suits our lifestyle and fitness level, we can increase our metabolic rate, burn calories, and build lean muscle mass. This not only aids in weight loss but also helps to maintain a healthy body composition.

4. Sleep and Recovery

Quality sleep is essential for overall health and well-being, including weight management. Research has shown that inadequate sleep can disrupt hormonal balance, leading to increased appetite and cravings for high-calorie foods. By prioritizing regular and restful sleep, we can support our weight loss efforts by maintaining a healthy hormonal balance.

2.1.3 Sustainable Weight Management with Frequency

While frequency can play a significant role in weight loss, it is important to approach weight management holistically and sustainably. Here are some frequency-based strategies for long-term weight management:

1. Consistency is Key

Consistency is crucial when it comes to weight management. By adopting healthy habits and maintaining them over time, we can create a sustainable lifestyle that supports our weight goals. This includes consistent meal patterns, regular exercise routines, and prioritizing restful sleep.

2. Mind-Body Connection

Developing a strong mind-body connection can help us better understand our body's needs and respond to them appropriately. By tuning in to our body's signals and practicing self-awareness, we can make informed choices about our nutrition, exercise, and overall well-being.

3. Self-Care and Stress Management

Stress can have a significant impact on weight management. Chronic stress can disrupt hormonal balance, leading to weight gain and difficulty in losing weight. Incorporating stress management techniques such as meditation, deep breathing exercises, and regular self-care practices can help reduce stress levels and support weight management efforts.

4. Seeking Professional Guidance

If weight management becomes challenging or overwhelming, seeking guidance from a healthcare professional or registered dietitian can provide valuable support. They can help create a personalized plan that takes into account individual needs, preferences, and goals, while also considering the role of frequency in weight management.

In conclusion, frequency plays a role in weight management by influencing metabolism, meal patterns, exercise routines, and sleep quality. By understanding and harnessing the power of frequency, we can optimize our weight loss efforts and achieve sustainable weight management.

2.2 How Frequency Affects Metabolism

Metabolism is a complex process that involves the conversion of food into energy and the regulation of various bodily functions. It plays a crucial role in weight management and overall health. The impact of frequency on metabolism is a fascinating area of study that has gained significant attention in recent years. In this section, we will explore how frequency affects metabolism and its implications for weight loss.

The Role of Frequency in Metabolism

Metabolism is influenced by various factors, including genetics, age, body composition, and lifestyle choices. However, emerging research suggests that frequency also plays a significant role in metabolic regulation. Frequency refers to the number of times an event occurs within a given time frame. In the context of metabolism, it refers to the frequency of meals and the timing of food intake.

The Effect of Meal Frequency on Metabolism

Traditionally, it was believed that consuming three meals a day was the optimal approach for maintaining a healthy metabolism. However, recent studies have challenged this notion and suggested that meal frequency may have a more significant impact on metabolism than previously thought.

Research has shown that increasing meal frequency to five or six smaller meals throughout the day can have several positive effects on metabolism. Firstly, it can help regulate blood sugar levels by preventing large spikes and crashes. This is particularly beneficial for individuals with diabetes or insulin resistance. Secondly, frequent meals can increase thermogenesis, the process by which the body burns calories to produce heat. This can lead to a higher overall energy expenditure and potentially aid in weight loss efforts.

On the other hand, some studies have also suggested that intermittent fasting, which involves reducing meal frequency and extending the fasting period, can have metabolic benefits. Intermittent fasting has been shown to improve insulin sensitivity, increase fat burning, and promote cellular repair processes. However, more research is needed to fully understand the long-term effects of intermittent fasting on metabolism.

The Timing of Food Intake and Metabolism

In addition to meal frequency, the timing of food intake also plays a role in metabolic regulation. The body's internal clock, known as the circadian rhythm, influences various physiological processes, including metabolism. Disruptions to the circadian rhythm, such as irregular sleep patterns or shift work, can have detrimental effects on metabolism and overall health.

Research has shown that consuming most of our calories earlier in the day, when our metabolism is naturally more active, can have metabolic benefits. This approach, known as time-restricted feeding, involves restricting food intake to a specific window of time, typically 8-10 hours, and fasting for the remaining hours of the day. Time-restricted feeding has been shown to improve insulin sensitivity, reduce inflammation, and promote weight loss.

The Impact of Frequency on Hormones

Hormones play a crucial role in metabolic regulation, and frequency can influence their secretion and function. For example, frequent meals can help stabilize blood sugar levels and prevent excessive insulin release, which is important for maintaining a healthy metabolism. Additionally, meal frequency can affect the release of other hormones involved in appetite regulation, such as ghrelin and leptin.

Ghrelin is known as the "hunger hormone" and stimulates appetite, while leptin is known as the "satiety hormone" and signals fullness. Research has shown that frequent meals can help regulate the secretion of these hormones, leading to better appetite control and potentially aiding in weight management.

The Importance of Nutrient Timing

While frequency plays a role in metabolism, it is essential to consider the quality and composition of the meals as well. Nutrient timing, or the strategic timing of macronutrient intake, can have a significant impact on metabolism and weight loss.

For example, consuming a balanced meal that includes protein, carbohydrates, and healthy fats within 30 minutes to an hour after exercise can enhance muscle recovery and promote muscle growth. This is because the body is more receptive to nutrient uptake during this post-workout period, known as the "anabolic window."

Similarly, consuming a protein-rich breakfast can help kickstart metabolism and provide sustained energy throughout the day. Protein requires more energy to digest and metabolize compared to carbohydrates and fats, which can increase calorie expenditure and potentially aid in weight loss.

Conclusion

Frequency plays a crucial role in metabolic regulation and can have significant implications for weight loss and overall health. Increasing meal frequency or adopting intermittent fasting approaches can impact metabolism and energy expenditure. Additionally, the timing of food intake and nutrient composition can further influence metabolic processes. Understanding the role of frequency in metabolism can empower individuals to make informed choices about their eating patterns and optimize their weight loss efforts.

2.3 Using Frequency to Enhance Weight Loss Efforts

Weight loss is a common goal for many individuals, and the journey to achieving and maintaining a healthy weight can often be challenging. While there are various factors that contribute to weight loss, such as diet and exercise, the role of frequency in enhancing weight loss efforts is an emerging area of interest. In this section, we will explore how frequency can be utilized to optimize weight loss and support sustainable weight management.

2.3.1 Understanding the Role of Frequency in Weight Loss

Frequency, in the context of weight loss, refers to the rate at which something occurs or repeats. In this case, it relates to the patterns and rhythms within the body that influence metabolism, energy expenditure, and fat burning. By understanding and harnessing the power of frequency, individuals can potentially enhance their weight loss journey.

2.3.2 The Impact of Frequency on Metabolism

Metabolism plays a crucial role in weight management, as it determines how efficiently the body converts food into energy. The rate at which our metabolism functions can be influenced by various factors, including genetics, age, and lifestyle choices. Interestingly, frequency has been found to have a significant impact on metabolism.

Research suggests that certain frequencies can stimulate the body's metabolic processes, leading to increased energy expenditure and fat burning. For example, studies have shown that exposure to specific sound frequencies can activate brown adipose tissue, a type of fat that helps burn calories to generate heat. By incorporating frequency-based techniques, individuals may be able to boost their metabolism and enhance their weight loss efforts.

2.3.3 Utilizing Frequency for Weight Loss Enhancement

1. **Frequency-based Exercise**: Incorporating frequency-based exercise routines into your weight loss regimen can be an effective strategy. High-intensity interval training (HIIT), for instance, involves alternating between short bursts of intense exercise and periods of rest or lower intensity. This approach not only challenges the body but also utilizes frequency to optimize calorie burning and fat loss.

2. **Frequency-based Dietary Approaches**: The timing and frequency of meals can also impact weight loss. Intermittent fasting, for example, involves cycling between periods of eating and fasting. This approach can help regulate insulin levels, improve metabolic flexibility, and promote fat burning. By aligning meal frequency with the body's natural rhythms, individuals can potentially enhance their weight loss efforts.

3. **Frequency-based Mindful Eating**: Mindful eating involves paying attention to the frequency and quality of food consumption. By practicing mindful eating techniques, such as savoring each bite, chewing slowly, and listening to the body's hunger and fullness cues, individuals can develop a healthier relationship with food. This approach can help prevent overeating and promote weight loss.

4. **Frequency-based Sleep Optimization**: Adequate sleep is essential for weight management. Research has shown that insufficient sleep can disrupt hormonal balance, leading to increased appetite and cravings for high-calorie foods. By prioritizing quality sleep and incorporating frequency-based sleep optimization techniques, such as relaxation exercises or soothing sounds, individuals can support their weight loss efforts.

2.3.4 Strategies for Sustainable Weight Management

While frequency can enhance weight loss efforts, it is important to adopt sustainable strategies for long-term weight management. Here are some frequency-based approaches that can support sustainable weight management:

1. **Consistency**: Consistency is key when it comes to weight management. By establishing a regular routine that incorporates frequency-based practices, individuals can create sustainable habits that support their weight loss goals.

2. **Self-awareness**: Developing self-awareness around eating patterns, emotional triggers, and stress levels can help individuals make conscious choices that support their weight management journey. By using frequency-based techniques, such as mindfulness and stress reduction practices, individuals can cultivate a deeper understanding of their body's needs and make healthier choices.

3. **Holistic Approach**: Weight management is not solely about diet and exercise. Taking a holistic approach that considers the interplay between frequency, sleep, stress, and overall well-being is essential. By addressing all aspects of health and incorporating frequency-based practices into various areas of life, individuals can optimize their weight loss efforts and maintain a healthy weight in the long run.

In conclusion, frequency can be a powerful tool in enhancing weight loss efforts. By understanding the impact of frequency on metabolism, utilizing frequency-based techniques, and adopting sustainable strategies, individuals can optimize their weight loss journey and achieve long-term weight management. Incorporating frequency into daily life can not only support weight loss but also contribute to overall well-being and a healthier lifestyle.

2.4 Frequency-Based Strategies for Sustainable Weight Management

Maintaining a healthy weight is a goal that many individuals strive to achieve. However, the journey towards sustainable weight management can often be challenging and frustrating. Fortunately, emerging research suggests that frequency-based strategies can play a significant role in supporting weight loss efforts and promoting long-term weight maintenance. In this section, we will explore various frequency-based strategies that can help individuals achieve sustainable weight management.

2.4.1 Harnessing the Power of Frequency for Weight Loss

When it comes to weight loss, the concept of frequency extends beyond the realm of diet and exercise. It encompasses the idea that the body operates on various frequencies, and by aligning these frequencies, individuals can optimize their weight loss journey. Here are some frequency-based strategies that can be incorporated into a weight loss plan:

2.4.1.1 Mindful Eating

One effective strategy for sustainable weight management is practicing mindful eating. Mindful eating involves paying close attention to the body's hunger and fullness cues, as well as the sensory experience of eating. By slowing down and savoring each bite, individuals can enhance their connection with their body's natural frequency and promote healthier eating habits. This practice can help prevent overeating and promote a more balanced approach to food consumption.

2.4.1.2 Frequency-Based Meal Timing

Another frequency-based strategy for weight management is meal timing. Research suggests that aligning meal times with the body's natural circadian rhythm can have a positive impact on weight loss. By consuming the majority of calories earlier in the day and gradually reducing intake in the evening, individuals can optimize their metabolism and improve weight loss outcomes. This approach takes into account the body's natural frequency patterns and supports its natural processes.

2.4.1.3 Incorporating High-Frequency Foods

Certain foods have a higher frequency than others, meaning they contain more vital energy and nutrients. Incorporating high-frequency foods into a weight loss plan can provide the body with the necessary nourishment while supporting weight loss efforts. These foods include fresh fruits and vegetables, whole grains, lean proteins, and healthy fats. By focusing on nutrient-dense options, individuals can optimize their overall health and well-being while working towards their weight management goals.

2.4.2 Balancing Frequency for Long-Term Weight Maintenance

Sustainable weight management is not just about losing weight; it also involves maintaining a healthy weight over the long term. Frequency-based strategies can play a crucial role in achieving this balance. Here are some strategies that can help individuals maintain their weight loss achievements:

2.4.2.1 Regular Physical Activity

Engaging in regular physical activity is essential for maintaining weight loss and overall well-being. Exercise helps to increase the body's frequency, promoting calorie expenditure and supporting muscle maintenance. By incorporating a combination of cardiovascular exercise, strength training, and flexibility exercises into their routine, individuals can optimize their weight management efforts and sustain their desired weight.

2.4.2.2 Mind-Body Practices

Mind-body practices such as yoga, meditation, and tai chi can be valuable tools for maintaining weight loss. These practices help individuals connect with their body's natural frequency, reduce stress, and promote a sense of overall well-being. By incorporating these practices into their daily routine, individuals can cultivate a positive mindset and develop a deeper understanding of their body's needs, leading to sustainable weight management.

2.4.2.3 Consistency and Self-Care

Consistency is key when it comes to maintaining weight loss. Establishing healthy habits and routines that align with the body's natural frequency can support long-term weight management. This includes prioritizing self-care activities such as getting enough sleep, managing stress levels, and practicing relaxation techniques. By taking care of their overall well-being, individuals can create a harmonious environment for their body to thrive and maintain a healthy weight.

2.4.3 The Holistic Approach to Weight Management

Frequency-based strategies for weight management go beyond traditional diet and exercise approaches. They encompass a holistic approach that considers the interconnectedness of the mind, body, and spirit. By incorporating these strategies into their daily lives, individuals can achieve sustainable weight management while promoting overall well-being. It is important to remember that each person's journey is unique, and finding the right balance of frequency-based strategies may require some experimentation and self-discovery.

In the next chapter, we will explore the link between frequency and stress management techniques, and how they can contribute to overall well-being.

3 Frequency and Stress

3.1 Understanding the Link Between Frequency and Stress

Stress has become an increasingly prevalent issue in today's fast-paced and demanding world. It affects individuals of all ages and can have a significant impact on overall well-being. In recent years, researchers have started to explore the relationship between frequency and stress, uncovering fascinating insights into how different frequencies can influence our stress levels and our ability to manage and cope with stress.

The Impact of Frequency on Stress Levels

Frequency, in the context of stress, refers to the vibrational energy that is emitted by various stimuli, such as sound, light, and even thoughts. Every object, including our bodies, emits a unique frequency. These frequencies can have a profound effect on our physiological and psychological states, including our stress levels.

Research has shown that certain frequencies have the ability to induce a state of relaxation and calmness, while others can stimulate the release of stress hormones and increase feelings of anxiety and tension. For example, low-frequency sounds, such as the gentle rustling of leaves or the sound of ocean waves, have been found to promote relaxation and reduce stress levels. On the other hand, high-frequency sounds, like loud sirens or screeching noises, can trigger a stress response in the body.

Similarly, visual stimuli can also impact stress levels. Studies have found that exposure to natural scenes, such as lush green landscapes or serene water bodies, can have a soothing effect on the mind and reduce stress. In contrast, environments with harsh lighting or chaotic visuals can contribute to feelings of stress and unease.

Managing Stress through Frequency Techniques

Understanding the influence of frequency on stress levels opens up a range of possibilities for managing and reducing stress. By intentionally exposing ourselves to frequencies that promote relaxation and calmness, we can effectively counteract the negative effects of stress on our well-being.

One popular technique for managing stress through frequency is known as binaural beats. Binaural beats involve listening to two slightly different frequencies in each ear, which creates a third frequency in the brain. This third frequency corresponds to a specific brainwave state, such as alpha or theta, which are associated with relaxation and reduced stress. By listening to binaural beats, individuals can entrain their brainwaves to a desired frequency and experience a state of deep relaxation.

Another technique that harnesses the power of frequency for stress reduction is known as sound therapy. Sound therapy involves using specific frequencies and vibrations to promote relaxation and balance in the body. This can be achieved through the use of instruments like singing bowls, tuning forks, or even specialized sound therapy devices. The vibrations produced by these instruments can help release tension, reduce stress, and restore harmony to the body and mind.

Using Frequency to Promote Relaxation and Well-Being

In addition to specific techniques like binaural beats and sound therapy, incorporating frequency into our daily lives can have a profound impact on our overall well-being. By consciously choosing to surround ourselves with frequencies that promote relaxation and calmness, we can create a more harmonious and stress-free environment.

One way to do this is by incorporating soothing sounds into our daily routines. This can involve listening to calming music, nature sounds, or even guided meditations that utilize specific frequencies known to induce relaxation. By incorporating these sounds into our daily lives, whether during work, leisure, or sleep, we can create a more peaceful and stress-free atmosphere.

Another way to harness the power of frequency for relaxation is through the use of visual stimuli. Surrounding ourselves with images and scenes that evoke a sense of tranquility, such as nature photographs or artwork depicting serene landscapes, can help create a visually calming environment. Additionally, incorporating soft and warm lighting can further enhance the relaxation response and reduce stress levels.

Frequency-Based Stress Reduction Strategies

In addition to incorporating frequency into our daily lives, there are several other strategies that can be employed to reduce stress and promote well-being. These strategies leverage the power of frequency to create a more balanced and harmonious state of being:

1. Mindfulness and meditation: Practicing mindfulness and meditation can help cultivate a state of present moment awareness and reduce stress. By focusing on the breath or a specific frequency, individuals can quiet the mind and promote relaxation.
2. Breathing exercises: Deep breathing exercises, such as diaphragmatic breathing or alternate nostril breathing, can help activate the body's relaxation response and reduce stress levels.
3. Yoga and tai chi: These ancient practices incorporate movement, breathwork, and mindfulness to promote relaxation, reduce stress, and enhance overall well-being.
4. Aromatherapy: Certain essential oils, such as lavender or chamomile, have been found to have calming properties. Incorporating these oils into a relaxation routine can help reduce stress and promote a sense of well-being.

By incorporating these frequency-based stress reduction strategies into our daily lives, we can effectively manage and reduce stress levels, leading to improved overall well-being and a greater sense of calmness and balance.

3.2 The Impact of Frequency on Stress Levels

Stress has become a prevalent issue in today's fast-paced and demanding world. It affects individuals of all ages and can have detrimental effects on both physical and mental well-being. However, recent research has shed light on the potential role of frequency in managing and reducing stress levels. In this section, we will explore the impact of frequency on stress and discuss various techniques that can be used to harness its benefits.

The Link Between Frequency and Stress

Before delving into the impact of frequency on stress levels, it is important to understand the underlying connection between the two. Every individual has a unique vibrational frequency, which can be influenced by various factors such as thoughts, emotions, and external stimuli. When we experience stress, our vibrational frequency tends to decrease, leading to a state of imbalance and disharmony within the body.

Frequency's Influence on Stress Levels

Research suggests that exposure to certain frequencies can help restore balance and reduce stress levels. Different frequencies have been found to have varying effects on the body and mind. For instance, low-frequency sounds, such as those produced by nature or calming music, have been shown to promote relaxation and induce a state of calmness. On the other hand, high-frequency sounds, such as those found in upbeat music, can stimulate the release of endorphins and elevate mood.

Moreover, studies have demonstrated that specific frequencies, such as those used in sound therapy or binaural beats, can help synchronize brainwaves and promote a state of deep relaxation. This synchronization of brainwaves has been associated with reduced stress, anxiety, and improved overall well-being.

Managing Stress through Frequency Techniques

There are several techniques that individuals can incorporate into their daily lives to manage stress levels using frequency. One such technique is mindfulness meditation, which involves focusing on the present moment and observing thoughts and sensations without judgment. By incorporating calming frequencies, such as those found in nature sounds or meditation music, individuals can enhance the effectiveness of their mindfulness practice and promote a sense of tranquility.

Another technique that has gained popularity is the use of frequency-based devices, such as biofeedback machines or wearable devices that emit specific frequencies. These devices work by detecting stress levels in the body and providing feedback in the form of auditory or visual cues. By using these devices regularly, individuals can become more aware of their stress levels and take proactive steps to reduce them.

Using Frequency to Promote Relaxation and Well-Being

In addition to managing stress, frequency can also be used to promote relaxation and overall well-being. By incorporating calming frequencies into daily routines, individuals can create a peaceful environment that supports relaxation and rejuvenation. This can be achieved through activities such as listening to soothing music, practicing yoga or tai chi, or engaging in guided imagery exercises.

Furthermore, research has shown that exposure to specific frequencies can have a positive impact on the body's stress response system. For example, studies have found that certain frequencies can help regulate cortisol levels, a hormone associated with stress. By modulating cortisol levels, frequency-based techniques can help individuals better cope with stress and maintain a state of balance.

Frequency-Based Stress Reduction Strategies

There are various frequency-based stress reduction strategies that individuals can incorporate into their daily lives. These strategies aim to harness the power of frequency to promote relaxation and reduce stress levels. Some of these strategies include:

1. Sound Therapy: Listening to calming music or nature sounds that have been specifically designed to induce relaxation and reduce stress.
2. Binaural Beats: Using headphones to listen to two different frequencies in each ear, which creates a third frequency that can synchronize brainwaves and promote relaxation.
3. Aromatherapy: Using essential oils with specific frequencies, such as lavender or chamomile, to create a calming atmosphere and promote relaxation.
4. Frequency-Based Breathing Techniques: Practicing deep breathing exercises while focusing on specific frequencies, which can help calm the mind and reduce stress.

By incorporating these strategies into daily routines, individuals can tap into the power of frequency to manage stress effectively and enhance overall well-being.

In conclusion, frequency has a profound impact on stress levels and can be used as a powerful tool for stress management. By understanding the link between frequency and stress, individuals can incorporate various frequency-based techniques into their daily lives to promote relaxation, reduce stress levels, and enhance overall well-being.

3.3 Managing Stress through Frequency Techniques

Stress has become a prevalent issue in today's fast-paced and demanding world. It can have a significant impact on our physical and mental well-being, affecting our sleep, weight, and overall quality of life. However, by understanding and harnessing the power of frequency, we can effectively manage and reduce stress levels, promoting relaxation and well-being.

3.3.1 The Link Between Frequency and Stress

Before delving into the techniques for managing stress through frequency, it is essential to understand the link between frequency and stress. Everything in the universe, including our bodies, is made up of energy vibrating at different frequencies. When we experience stress, our body's energy becomes imbalanced, leading to physical and emotional tension.

3.3.2 The Impact of Frequency on Stress Levels

Research has shown that specific frequencies can have a profound impact on our stress levels. For instance, low-frequency sounds, such as the gentle hum of nature or calming music, can induce a state of relaxation and reduce stress. On the other hand, high-frequency sounds, like loud noises or chaotic environments, can increase stress levels and trigger anxiety.

3.3.3 Frequency Techniques for Stress Management

1. **Binaural Beats**: Binaural beats are a powerful tool for managing stress. By listening to two slightly different frequencies in each ear, the brain creates a third frequency, known as the binaural beat. These

beats can synchronize brainwaves, promoting a state of deep relaxation and reducing stress.

2. **Guided Imagery**: Guided imagery involves using the power of visualization to create a calming and stress-free mental environment. By imagining peaceful scenes or engaging in positive visualizations, we can shift our focus away from stressors and induce a state of relaxation.

3. **Breathing Exercises**: Deep breathing exercises have long been recognized as an effective stress management technique. By focusing on slow, deep breaths, we activate the body's relaxation response, reducing stress hormones and promoting a sense of calm.

4. **Meditation**: Meditation is a practice that involves focusing the mind and achieving a state of mental clarity and calm. By incorporating frequency-based meditation techniques, such as chanting or using specific sound frequencies, we can enhance the stress-reducing benefits of meditation.

3.3.4 Frequency-Based Relaxation Methods

1. **Sound Therapy**: Sound therapy utilizes specific frequencies and vibrations to promote relaxation and reduce stress. This can be achieved through the use of singing bowls, tuning forks, or even specialized sound therapy apps that emit calming frequencies.

2. **Music Therapy**: Music has a profound impact on our emotions and can be a powerful tool for stress reduction. By listening to soothing music with specific frequencies, such as classical music or nature sounds, we can create a peaceful atmosphere and promote relaxation.

3. **Nature Therapy**: Spending time in nature has been shown to have numerous benefits for stress reduction. The natural frequencies and sounds of the environment can help restore balance to our energy and promote a sense of calm and well-being.

3.3.5 Frequency-Based Stress Reduction Strategies

1. **Daily Frequency Practice**: Incorporating frequency-based stress reduction techniques into our daily routine can have long-lasting benefits. Setting aside dedicated time each day for practices such as meditation, sound therapy, or guided imagery can help us manage stress more effectively.

2. **Creating a Stress-Free Environment**: Our physical environment plays a significant role in our stress levels. By creating a calm and peaceful space at home or work, we can minimize external stressors and create a supportive environment for relaxation.

3. **Self-Care and Mindfulness**: Engaging in self-care activities, such as taking a warm bath, practicing yoga, or engaging in hobbies, can help reduce stress. Additionally, cultivating mindfulness and being present in the moment can help us manage stress more effectively.

4. **Seeking Support**: If stress becomes overwhelming, it is essential to seek support from professionals or support groups. Therapists, counselors, or stress management programs can provide guidance and techniques tailored to individual needs.

By incorporating these frequency-based stress management techniques into our lives, we can effectively reduce stress levels, promote relaxation, and enhance our overall well-being. It is important to remember that managing stress is a continuous process, and finding the right combination of techniques that work for each individual is key. With dedication and practice, we can harness the power of frequency to lead a more balanced and stress-free life.

3.4 Using Frequency to Promote Relaxation and Well-Being

In our fast-paced and stressful modern lives, finding ways to relax and promote overall well-being is essential. One powerful tool that can be utilized for this purpose is frequency. Frequency refers to the rate at which something vibrates or oscillates, and it has been found to have a profound impact on our physical, mental, and emotional states. By understanding and harnessing the power of frequency, we can effectively promote relaxation and enhance our overall well-being.

3.4.1 The Science Behind Frequency and Relaxation

To understand how frequency can promote relaxation, it is important to delve into the science behind it. Every object in the universe, including our bodies, has a natural frequency at which it vibrates. When we are exposed to external frequencies that align with our natural frequency, it can have a harmonizing effect on our body and mind.

Research has shown that certain frequencies have the ability to stimulate the production of brainwave patterns associated with relaxation and calmness. For example, low-frequency delta waves (0.5-4 Hz) are associated with deep sleep and relaxation, while alpha waves (8-12 Hz) are linked to a relaxed and meditative state. By exposing ourselves to these frequencies, either through sound or other means, we can induce a state of relaxation and promote overall well-being.

3.4.2 Frequency-Based Techniques for Relaxation

There are various techniques that can be used to harness the power of frequency and promote relaxation. One popular method is through the use of binaural beats. Binaural beats are created by playing two slightly different frequencies in each ear, which then interact in the brain to produce a perceived beat frequency. This beat frequency corresponds to a specific brainwave pattern, such as alpha or theta waves, inducing a state of relaxation.

Another technique is the use of isochronic tones. Isochronic tones are regular beats of a single tone that are turned on and off rapidly. These tones are highly effective in entraining the brain to a desired frequency, promoting relaxation and reducing stress.

Sound therapy, such as listening to calming music or nature sounds, is another frequency-based technique that can promote relaxation. Certain types of music, such as classical or ambient music, have been found to have a soothing effect on the mind and body, helping to induce a state of relaxation.

3.4.3 The Benefits of Frequency-Based Relaxation

Using frequency to promote relaxation and well-being offers numerous benefits. Firstly, it can help to reduce stress and anxiety. By entraining the brain to a relaxed state, frequency-based techniques can calm the mind and release tension, allowing individuals to experience a greater sense of peace and tranquility.

Additionally, frequency-based relaxation techniques can improve sleep quality. As mentioned earlier, certain frequencies, such as delta waves, are associated with deep sleep. By listening to delta wave frequencies before bed, individuals can enhance their sleep and wake up feeling more refreshed and rejuvenated.

Furthermore, frequency-based relaxation techniques have been found to have a positive impact on mental health. They can help to alleviate symptoms of depression and anxiety, improve mood, and enhance overall emotional well-being. By promoting relaxation and reducing stress, frequency-based practices can contribute to a more positive mindset and improved mental resilience.

3.4.4 Incorporating Frequency-Based Relaxation into Daily Life

To incorporate frequency-based relaxation techniques into daily life, it is important to establish a regular practice. This can be done by setting aside dedicated time each day for relaxation exercises, such as listening to binaural beats or isochronic tones, engaging in sound therapy, or practicing mindfulness meditation.

Creating a calm and peaceful environment is also crucial for effective relaxation. This can be achieved by minimizing distractions, dimming lights, and creating a comfortable space where you can fully immerse yourself in the frequency-based practice.

It is important to note that frequency-based relaxation techniques may not work for everyone in the same way. Each individual has a unique response to different frequencies, so it may require some experimentation to find the techniques that work best for you. Additionally, it is advisable to consult with a healthcare professional before incorporating any new relaxation practices, especially if you have any underlying health conditions.

In conclusion, frequency has a profound impact on our relaxation and overall well-being. By understanding the science behind frequency and utilizing frequency-based techniques, we can effectively promote relaxation, reduce stress, and enhance our mental and emotional well-being. Incorporating these practices into our daily lives can lead to a greater sense of calmness, improved sleep, and a more positive outlook on life.

3.5 Frequency-Based Stress Reduction Strategies

Stress has become an increasingly prevalent issue in today's fast-paced and demanding world. The constant pressure and demands placed on individuals can have detrimental effects on their physical and mental well-being. However, emerging research suggests that frequency-based strategies can be effective in reducing stress and promoting relaxation. In this section, we will explore various frequency-based stress reduction strategies that can help individuals manage and alleviate stress in their lives.

3.5.1 Binaural Beats for Stress Reduction

One popular frequency-based technique for stress reduction is the use of binaural beats. Binaural beats are created by playing two slightly different frequencies in each ear, which then interact in the brain to produce a perceived beat frequency. This beat frequency has been found to have a calming effect on the mind and body, helping to reduce stress and promote relaxation.

Research has shown that listening to binaural beats can lead to a decrease in cortisol levels, the hormone associated with stress. By reducing cortisol levels, binaural beats can help individuals feel more relaxed and less anxious. They can also improve sleep quality, which is often disrupted by stress.

3.5.2 Solfeggio Frequencies for Stress Relief

Solfeggio frequencies are a set of ancient musical frequencies that are believed to have healing properties. These frequencies are said to resonate with the body's energy centers, or chakras, and can help restore balance and harmony. When it comes to stress reduction, specific solfeggio frequencies are often used.

For example, the 528 Hz frequency, also known as the "Love Frequency," is believed to promote feelings of peace and reduce stress. Listening to music or tones that incorporate this frequency can help individuals relax and unwind. Similarly, the 432 Hz frequency is associated with a sense of calm and tranquility, making it another effective tool for stress reduction.

3.5.3 Guided Meditation with Frequency-Based Soundscapes

Guided meditation is a popular practice for stress reduction, and when combined with frequency-based soundscapes, it can be even more effective. Guided meditation involves listening to a recorded voice that provides instructions and guidance for relaxation and stress relief. By incorporating frequency-based soundscapes into guided meditation sessions, individuals can enhance the calming effects and deepen their state of relaxation.

Frequency-based soundscapes often include soothing nature sounds, such as gentle waves or chirping birds, combined with specific frequencies that promote relaxation. These soundscapes can help individuals focus their attention, quiet their minds, and let go of stress and tension.

3.5.4 Sound Bath Therapy for Stress Reduction

Sound bath therapy is a practice that involves immersing oneself in the healing vibrations of various instruments, such as singing bowls, gongs, and tuning forks. These instruments produce specific frequencies that resonate with the body and promote relaxation. Sound bath therapy can be a powerful tool for stress reduction, as it helps to shift the mind and body into a state of deep relaxation and calm.

During a sound bath session, individuals lie down and allow the vibrations and frequencies to wash over them. The soothing sounds and vibrations can help release tension, reduce anxiety, and promote a sense of well-being. Sound bath therapy is often used as a complementary approach to traditional stress management techniques and can be particularly beneficial for individuals who have difficulty quieting their minds or find it challenging to relax.

3.5.5 Frequency-Based Breathing Techniques

Breathing techniques have long been used as a way to reduce stress and promote relaxation. When combined with frequency-based techniques, breathing exercises can be even more effective in managing stress. One such technique is known as resonant breathing, which involves breathing at a specific frequency to synchronize the heart rate and promote a state of calm.

Resonant breathing typically involves inhaling and exhaling at a rate of around six breaths per minute, which corresponds to a frequency of approximately 0.1 Hz. This frequency is believed to activate the body's relaxation response and help regulate the autonomic nervous system. By practicing resonant breathing with the guidance of frequency-based audio cues, individuals can enhance the stress-reducing effects and experience a greater sense of calm and well-being.

3.5.6 Incorporating Frequency-Based Activities into Daily Life

In addition to specific techniques, incorporating frequency-based activities into daily life can also help reduce stress and promote overall well-being. This can include listening to calming music or nature sounds with embedded frequencies, using frequency-based apps or devices for relaxation, or even spending time in nature and immersing oneself in the natural frequencies of the environment.

By integrating these activities into daily routines, individuals can create a more balanced and stress-free lifestyle. It is important to note that everyone may respond differently to frequency-based stress reduction strategies, so it is essential to find the techniques that resonate most with you and incorporate them into your own unique self-care routine.

In conclusion, frequency-based stress reduction strategies offer a promising approach to managing and alleviating stress. Whether through the use of binaural beats, solfeggio frequencies, guided meditation, sound bath therapy, breathing techniques, or incorporating frequency-based activities into daily life, individuals have a range of options to choose from. By exploring and experimenting with these strategies, individuals can discover the techniques that work best for them and experience the profound benefits of frequency in reducing stress and promoting overall well-being.

3.6 The Role of Frequency in Stress Resilience

Stress is an inevitable part of life, and its impact on our well-being cannot be underestimated. The constant demands and pressures we face can take a toll on our physical and mental health. However, recent research has shed light on the role of frequency in stress resilience and how it can help us better cope with and manage stress.

3.6.1 Understanding the Link Between Frequency and Stress

To comprehend the role of frequency in stress resilience, it is essential to understand the connection between frequency and our body's stress response. Every individual has a unique vibrational frequency, which can be influenced by various factors such as thoughts, emotions, and external stimuli. When we experience stress, our frequency tends to become imbalanced, leading to a range of negative effects on our well-being.

3.6.2 The Impact of Frequency on Stress Levels

Research has shown that certain frequencies can have a profound impact on our stress levels. For instance, low-frequency vibrations, such as those emitted by calming music or nature sounds, have been found to induce a relaxation response in the body. These frequencies help reduce the production of stress hormones like cortisol and promote a sense of calm and tranquility.

Conversely, high-frequency vibrations, such as those associated with loud noises or chaotic environments, can trigger a stress response in the body. These frequencies can increase cortisol levels, elevate heart rate, and contribute to feelings of anxiety and tension.

3.6.3 Managing Stress through Frequency Techniques

Understanding the influence of frequency on stress levels opens up new possibilities for managing and reducing stress. By incorporating specific frequency techniques into our daily lives, we can enhance our stress resilience and improve our overall well-being.

One effective technique is the use of binaural beats, which involve listening to two slightly different frequencies in each ear. This creates a third frequency in the brain, known as the "beat frequency," which can help induce a state of relaxation and reduce stress. Binaural beats have been found to promote the production of alpha and theta brainwaves, which are associated with deep relaxation and a calm mental state.

Another technique is the practice of mindfulness meditation, which involves focusing on the present moment without judgment. Mindfulness meditation has been shown to reduce stress and anxiety by helping individuals become more aware of their thoughts and emotions. By incorporating specific frequency tones or music into mindfulness meditation sessions, individuals can further enhance the relaxation and stress-reducing benefits of this practice.

3.6.4 Using Frequency to Promote Relaxation and Well-Being

In addition to specific frequency techniques, incorporating relaxation practices that utilize frequency can have a significant impact on stress resilience. For example, practices such as sound therapy, where specific frequencies are used to create a harmonious and calming environment, can help reduce stress levels and promote a sense of well-being.

Sound baths, a form of sound therapy, involve immersing oneself in the soothing sounds and vibrations produced by various instruments, such as singing bowls or gongs. These vibrations can help release tension, promote relaxation, and restore balance to the body's energy systems.

Another effective practice is the use of frequency-based guided imagery or visualization exercises. By combining specific frequencies with guided imagery, individuals can create a mental landscape that promotes relaxation and reduces stress. This technique allows individuals to tap into their imagination and create a safe and peaceful space within their minds, helping them to better cope with and manage stress.

3.6.5 Frequency-Based Stress Reduction Strategies

In addition to specific techniques and practices, there are several frequency-based strategies that can be incorporated into daily life to reduce stress and enhance stress resilience. These strategies include:

1. Creating a calming environment: Surrounding oneself with soothing sounds and frequencies, such as nature sounds or ambient music, can help create a peaceful and stress-free environment.
2. Engaging in regular physical activity: Exercise has been shown to reduce stress and improve overall well-being. By incorporating activities such as yoga or tai chi, which often incorporate specific frequencies and vibrations, individuals can further enhance the stress-reducing benefits of physical activity.
3. Practicing deep breathing exercises: Deep breathing exercises, such as diaphragmatic breathing or box breathing, can help activate the body's relaxation response and reduce stress levels. By focusing on the breath and incorporating specific frequencies into these exercises, individuals can enhance their stress-reducing effects.
4. Incorporating frequency-based self-care practices: Engaging in self-care activities that incorporate specific frequencies, such as aromatherapy or massage therapy, can help reduce stress and promote relaxation.

By incorporating these frequency-based strategies into our daily lives, we can enhance our stress resilience and improve our overall well-being. The power of frequency in stress management cannot be underestimated, and by harnessing its potential, we can better navigate the challenges of life and cultivate a greater sense of peace and balance.

4 Exploring the Various Effects of Frequency

4.1 Frequency and Cognitive Function

Cognitive function refers to the mental processes and abilities that allow us to think, reason, learn, and remember. It encompasses various aspects such as attention, memory, problem-solving, decision-making, and creativity. The impact of frequency on cognitive function is a fascinating area of study that has gained significant attention in recent years.

4.1.1 The Role of Frequency in Brain Function

The brain operates on electrical impulses and relies on the synchronization of neural activity to function optimally. This synchronization is achieved through the coordination of different frequencies of brainwaves. Brainwaves are rhythmic patterns of electrical activity generated by the brain, and they can be categorized into different frequency bands, including delta, theta, alpha, beta, and gamma.

Each frequency band is associated with specific mental states and cognitive processes. For example, delta waves are dominant during deep sleep and unconsciousness, while theta waves are present during relaxation and meditation. Alpha waves are associated with a relaxed but alert state, and beta waves are linked to focused attention and active thinking. Gamma waves, the fastest brainwave frequency, are associated with higher cognitive functions such as memory consolidation and information processing.

4.1.2 Enhancing Cognitive Function with Frequency

Research has shown that specific frequencies of brainwave activity can enhance cognitive function in various ways. For example, alpha and theta frequencies have been found to promote relaxation, creativity, and problem-solving abilities. These frequencies can be induced through techniques such as meditation, deep breathing exercises, and listening to specific types of music.

Moreover, studies have demonstrated that exposure to certain frequencies, such as gamma waves, can improve memory and learning capabilities. Gamma wave stimulation has been shown to enhance synaptic plasticity, which is the brain's ability to form and strengthen connections between neurons. This can lead to improved information processing, attention, and overall cognitive performance.

4.1.3 Frequency and Attention

Attention is a crucial aspect of cognitive function, as it allows us to focus on relevant information while filtering out distractions. Research has shown that specific frequencies can influence attentional processes. For instance, alpha waves have been associated with improved attentional control and the ability to sustain focus. By training the brain to produce alpha waves through neurofeedback or other techniques, individuals may experience enhanced attention and concentration.

4.1.4 Frequency and Memory

Memory is another fundamental cognitive function that can be influenced by frequency. Studies have shown that certain frequencies, such as theta waves, are closely linked to memory consolidation and retrieval processes. Theta wave stimulation during sleep has been found to enhance memory formation and retention. Additionally, exposure to alpha and gamma frequencies has been associated with improved working memory and information processing speed.

4.1.5 Frequency-Based Cognitive Enhancement Techniques

Various techniques and technologies have been developed to harness the power of frequency for cognitive enhancement. One such technique is binaural beats, which involve listening to two slightly different frequencies in each ear to create a perceived third frequency in the brain. Binaural beats have been found to induce specific brainwave patterns and can be used to promote relaxation, focus, and creativity.

Another approach is transcranial magnetic stimulation (TMS), which uses magnetic fields to stimulate specific regions of the brain. TMS has shown promise in enhancing cognitive function, particularly in the areas of attention, memory, and problem-solving.

4.1.6 The Future of Frequency and Cognitive Enhancement

As our understanding of the brain and its complex relationship with frequency continues to evolve, the potential for cognitive enhancement through frequency-based interventions is vast. Ongoing research aims to uncover more precise methods of targeting specific brain regions and frequencies to optimize cognitive function.

It is important to note that while frequency-based techniques hold promise for cognitive enhancement, individual responses may vary. Factors such as baseline cognitive abilities, overall health, and personal preferences can influence the effectiveness of these interventions. Therefore, it is advisable to consult with healthcare professionals or experts in the field before incorporating frequency-based practices into one's routine.

In conclusion, the impact of frequency on cognitive function is a fascinating area of study. By understanding the role of different brainwave frequencies and utilizing frequency-based techniques, individuals may be able to enhance attention, memory, problem-solving abilities, and overall cognitive performance. As research in this field continues to advance, the potential for frequency-based cognitive enhancement holds great promise for improving the quality of our mental lives.

4.2 The Influence of Frequency on Mood and Emotions

Frequency, as a fundamental aspect of our existence, plays a significant role in shaping our mood and emotions. The vibrations and oscillations that occur at different frequencies have the power to impact our mental and emotional states in profound ways. In this section, we will explore how frequency influences our mood and emotions, and how we can harness its potential to enhance our overall well-being.

4.2.1 The Emotional Resonance of Frequency

Every emotion we experience is accompanied by a specific frequency pattern within our bodies. These frequencies can be influenced by external factors such as music, sounds, and even the electromagnetic waves that surround us. When we are exposed to frequencies that resonate with our emotional state, they have the potential to either amplify or alleviate our emotions.

For example, research has shown that certain frequencies can induce feelings of relaxation and calmness. These frequencies, often found in nature sounds like ocean waves or gentle rain, have a soothing effect on our nervous system, helping to reduce stress and anxiety. On the other hand, higher frequency sounds like upbeat music can stimulate feelings of happiness and excitement, boosting our mood and energy levels.

4.2.2 The Role of Frequency in Emotional Regulation

The influence of frequency on our mood and emotions goes beyond mere resonance. It also plays a crucial role in regulating our emotional states. By consciously exposing ourselves to specific frequencies, we can actively modulate our emotions and promote a more positive mindset.

One way frequency can impact emotional regulation is through its effect on brainwave activity. Different brainwave frequencies, such as alpha, beta, theta, and delta, are associated with different mental states and emotions. For instance, alpha waves are linked to relaxation and creativity, while beta waves are associated with focus and alertness. By entraining our brainwaves to specific frequencies through techniques like binaural beats or isochronic tones, we can shift our emotional state towards the desired outcome.

4.2.3 Frequency and Emotional Healing

Frequency also has the potential to facilitate emotional healing and release. Just as sound therapy has been used for centuries to promote physical healing, it can also be employed to address emotional imbalances and traumas. By exposing ourselves to specific frequencies that resonate with the emotions we wish to heal, we can initiate a process of emotional release and transformation.

For example, certain frequencies have been found to have a calming effect on the nervous system, helping to alleviate symptoms of anxiety, depression, and post-traumatic stress disorder (PTSD). These frequencies can be incorporated into therapeutic practices such as meditation, sound baths, or frequency-based therapies to support emotional healing and well-being.

4.2.4 Frequency and Emotional Resilience

In addition to influencing our immediate mood and emotions, frequency practices can also enhance our emotional resilience. By regularly engaging with frequencies that promote emotional balance and well-being, we can strengthen our ability to cope with stress, adversity, and negative emotions.

Research suggests that exposure to specific frequencies can increase the production of neurotransmitters such as serotonin and dopamine, which are associated with feelings of happiness and well-being. By cultivating a positive emotional state through frequency practices, we can build emotional resilience and improve our overall mental health.

4.2.5 Frequency-Based Strategies for Emotional Well-Being

Incorporating frequency-based strategies into our daily lives can have a profound impact on our mood and emotional well-being. Here are some practical techniques to harness the power of frequency for emotional enhancement:

1. **Sound Therapy**: Explore the use of sound therapy tools such as singing bowls, tuning forks, or frequency-specific music to create a harmonious environment that supports emotional balance.
2. **Meditation and Mindfulness**: Incorporate frequency-based meditation practices into your daily routine. Focus on breathing techniques and visualization exercises while listening to frequency-enhanced music or guided meditations.
3. **Nature Immersion**: Spend time in nature and immerse yourself in the natural frequencies of the environment. Take walks in the park, listen to the sounds of birds chirping, or simply sit by a flowing river to experience the calming effects of nature's frequencies.
4. **Frequency-Based Affirmations**: Create positive affirmations that resonate with the emotional state you wish to cultivate. Repeat these affirmations while listening to frequency-enhanced audios to reinforce the desired emotional state.
5. **Self-Reflection and Journaling**: Engage in self-reflection and journaling practices to explore and understand your emotions better. Use frequency-enhanced music or sounds as a backdrop to create a conducive environment for emotional exploration.

By incorporating these frequency-based strategies into our lives, we can tap into the transformative power of frequency to enhance our mood, regulate our emotions, and cultivate a greater sense of emotional well-being.

Conclusion

Frequency has a profound influence on our mood and emotions. By understanding and harnessing the power of frequency, we can actively shape our emotional states, promote emotional healing, and enhance our overall well-being. Through practices such as sound therapy, meditation, and nature immersion, we can create a harmonious environment that supports emotional balance and resilience. By incorporating frequency-based strategies into our daily lives, we can unlock the potential for emotional transformation and cultivate a positive and fulfilling emotional experience.

4.3 Frequency's Impact on Physical Performance

Physical performance is a crucial aspect of overall well-being, and the impact of frequency on physical performance cannot be overlooked. The frequency at which our body operates plays a significant role in determining our ability to perform physical tasks, whether it be athletic endeavors, daily activities, or even simple movements. In this section, we will explore how frequency influences physical performance and the various ways in which it can be optimized to enhance our capabilities.

4.3.1 The Role of Frequency in Energy Levels

One of the key factors that determine physical performance is the energy levels within our body. Frequency has a direct influence on our energy levels, as it affects the efficiency of our body's energy production processes. When our body operates at an optimal frequency, the energy production pathways function smoothly, resulting in higher energy levels and improved physical performance.

4.3.2 Enhancing Strength and Endurance

Frequency also plays a vital role in enhancing strength and endurance. When our body operates at an optimal frequency, it allows for efficient muscle contraction and relaxation, leading to improved strength and endurance capabilities. Additionally, frequency optimization can enhance the oxygen-carrying capacity of our blood, improving endurance and reducing fatigue during physical activities.

4.3.3 Speed and Agility

The impact of frequency on physical performance extends to speed and agility as well. When our body operates at an optimal frequency, it allows for faster neural communication, leading to improved reaction times and agility. This can be particularly beneficial for athletes participating in sports that require quick movements and rapid decision-making.

4.3.4 Coordination and Balance

Frequency also plays a crucial role in coordination and balance. When our body operates at an optimal frequency, it allows for better synchronization of muscle groups, leading to improved coordination and balance. This can be particularly beneficial for activities that require precise movements and control, such as dancing, gymnastics, or martial arts.

4.3.5 Injury Prevention and Recovery

Optimizing frequency can also contribute to injury prevention and recovery. When our body operates at an optimal frequency, it promotes better circulation, which aids in the delivery of nutrients and oxygen to the muscles and tissues. This, in turn, helps in preventing injuries and expediting the recovery process in case of any physical trauma.

4.3.6 Mental Focus and Concentration

Physical performance is not solely dependent on the body; it also relies on mental focus and concentration. Frequency optimization can enhance mental clarity, focus, and concentration, allowing individuals to perform at their best during physical activities. This can be particularly beneficial for athletes and individuals participating in competitive sports.

4.3.7 Frequency-Based Techniques for Physical Performance Enhancement

There are various frequency-based techniques that can be employed to enhance physical performance. These techniques include:

4.3.7.1 Binaural Beats

Binaural beats are a form of auditory stimulation that involves listening to two slightly different frequencies in each ear. This technique has been found to enhance focus, relaxation, and overall physical performance. By incorporating binaural beats into pre-workout or training routines, individuals can experience improved physical performance.

4.3.7.2 Frequency-Specific Music

Listening to frequency-specific music, such as those designed to stimulate specific brainwave patterns, can also have a positive impact on physical performance. Certain frequencies have been found to enhance focus, motivation, and energy levels, thereby improving physical performance during workouts or physical activities.

4.3.7.3 Biofeedback Training

Biofeedback training involves using electronic devices to monitor and provide feedback on various physiological parameters, such as heart rate, muscle tension, and brainwave activity. By utilizing biofeedback techniques, individuals can learn to regulate their body's frequency and optimize it for improved physical performance.

4.3.7.4 Frequency-Based Exercise Routines

Designing exercise routines that incorporate specific frequencies can also be beneficial for physical performance enhancement. For example, incorporating high-frequency movements or exercises that align with specific brainwave patterns can help individuals achieve optimal physical performance.

Conclusion

Frequency has a profound impact on physical performance, influencing energy levels, strength, endurance, speed, agility, coordination, balance, injury prevention, and mental focus. By understanding and optimizing the frequency at which our body operates, we can unlock our full physical potential and enhance our overall well-being. Incorporating frequency-based techniques, such as binaural beats, frequency-specific music, biofeedback training, and frequency-based exercise routines, can further amplify the positive effects on physical performance.

4.4 Frequency and Immune System Function

The immune system is a complex network of cells, tissues, and organs that work together to defend the body against harmful pathogens and foreign substances. It plays a crucial role in maintaining overall health and well-being. Recent research has shown that frequency can have a significant impact on immune system function, influencing its ability to fight off infections and diseases. In this section, we will explore the relationship between frequency and immune system function, and how optimizing frequency can enhance our body's defense mechanisms.

4.4.1 The Link Between Frequency and Immune System

The immune system is regulated by various factors, including genetics, lifestyle, and environmental influences. One emerging area of research is the role of frequency in modulating immune system function. Studies have shown that different frequencies can have distinct effects on immune cells, such as lymphocytes, macrophages, and natural killer cells. These cells play a vital role in identifying and eliminating pathogens, infected cells, and abnormal cells within the body.

4.4.2 Frequency's Impact on Immune System Response

Research suggests that specific frequencies can stimulate immune system activity, leading to a more robust and efficient response against infections and diseases. For example, studies have shown that exposure to certain frequencies can enhance the production of cytokines, which are signaling molecules that regulate immune cell communication and activation. This increased cytokine production can help to amplify the immune response and promote the elimination of pathogens.

Conversely, disruptions in frequency patterns, such as chronic stress or exposure to harmful electromagnetic frequencies, can impair immune system function. Stress, in particular, has been shown to suppress immune system activity, making individuals more susceptible to infections and diseases. By understanding the impact of frequency on immune system function, we can develop strategies to optimize frequency patterns and support a healthy immune response.

4.4.3 Optimizing Frequency for a Healthy Immune System

There are several ways in which we can optimize frequency to support a healthy immune system:

4.4.3.1 Frequency-Based Therapies

Certain frequency-based therapies, such as sound therapy and bioresonance therapy, have been shown to have positive effects on immune system function. These therapies involve the use of specific frequencies to stimulate immune cells and promote overall well-being. Sound therapy, for example, utilizes specific sound frequencies to induce a state of relaxation and balance within the body, which can enhance immune system function.

4.4.3.2 Lifestyle Factors

Lifestyle factors, such as diet, exercise, and stress management, can also influence immune system function. A healthy, balanced diet rich in nutrients and antioxidants can provide the necessary building blocks for a strong immune system. Regular exercise has been shown to enhance immune system activity and reduce the risk of chronic diseases. Additionally, managing stress through techniques like meditation, deep breathing, and mindfulness can help to regulate frequency patterns and support immune system function.

4.4.3.3 Sleep and Immune System Health

Sleep plays a vital role in immune system function. During sleep, the body undergoes essential processes that support immune system activity, such as the production of immune cells and the release of cytokines. Disruptions in sleep patterns, such as insufficient sleep or poor sleep quality, can impair immune system function and increase susceptibility to infections. Optimizing frequency patterns during sleep, such as using calming frequencies or white noise, can promote deep and restorative sleep, thereby supporting immune system health.

4.4.4 Frequency and Immune System Disorders

Research has also explored the potential of frequency-based therapies in managing immune system disorders. For example, studies have investigated the use of specific frequencies in the treatment of autoimmune diseases, such as rheumatoid arthritis and multiple sclerosis. While more research is needed in this area, preliminary findings suggest that frequency-based interventions may help modulate immune system activity and alleviate symptoms associated with these conditions.

4.4.5 Conclusion

Frequency plays a crucial role in immune system function and overall well-being. By understanding the impact of frequency on immune system activity, we can optimize our frequency patterns to support a healthy immune response. Incorporating frequency-based therapies, adopting a healthy lifestyle, prioritizing quality sleep, and managing stress can all contribute to enhancing immune system function. Further research in this field will continue to shed light on the intricate relationship between frequency and immune system health, paving the way for innovative approaches to immune system disorders and overall well-being.

5 The Benefits of Frequency

5.1 Enhancing Overall Well-Being with Frequency

Frequency, the measure of the number of occurrences of a repeating event per unit of time, has a profound impact on various aspects of our lives, including sleep, weight loss, stress, and overall well-being. In this section, we will explore how frequency can be harnessed to enhance our overall well-being and improve our quality of life.

5.1.1 Frequency's Role in Mental Health

Mental health is a crucial component of overall well-being, and frequency plays a significant role in maintaining and improving it. Research has shown that specific frequencies can have a profound impact on our brainwaves and mental states. For example, alpha waves, which have a frequency range of 8 to 12 Hz, are associated with a relaxed and calm state of mind. By exposing ourselves to alpha frequency through techniques like binaural beats or meditation, we can promote a sense of tranquility and reduce anxiety and stress.

Furthermore, frequency-based practices have been found to be effective in managing depression and mood disorders. Studies have shown that certain frequencies, such as gamma waves (30 to 100 Hz), can stimulate the release of endorphins and serotonin, which are neurotransmitters associated with happiness and well-being. By incorporating frequency techniques into our daily routine, we can potentially alleviate symptoms of depression and enhance our overall mental well-being.

5.1.2 Frequency and Physical Health Benefits

In addition to its impact on mental health, frequency also has significant effects on our physical well-being. Various frequencies have been found to influence physical performance and endurance. For instance, research suggests that exposure to beta waves (13 to 30 Hz) can increase alertness and improve concentration, making it beneficial for athletes and individuals engaging in physical activities. By incorporating frequency-based techniques into our exercise routine, we can potentially enhance our physical performance and achieve better results.

Moreover, frequency-based approaches have been explored for pain management and rehabilitation. Studies have shown that specific frequencies, such as delta waves (0.5 to 4 Hz), can stimulate the release of endogenous opioids, which are natural pain-relieving substances produced by the body. By utilizing frequency techniques, such as transcutaneous electrical nerve stimulation (TENS), individuals suffering from chronic pain or undergoing rehabilitation can potentially experience relief and accelerate the healing process.

5.1.3 The Holistic Benefits of Frequency Practices

One of the remarkable aspects of frequency is its ability to impact multiple dimensions of our well-being simultaneously. By incorporating frequency practices into our daily lives, we can experience holistic benefits that positively influence our mental, emotional, and physical states.

For example, frequency-based relaxation methods, such as guided meditation or sound therapy, can induce a state of deep relaxation and promote overall well-being. These practices help reduce stress, lower blood pressure, and improve sleep quality. By regularly engaging in relaxation techniques that utilize specific frequencies, we can create a harmonious balance within ourselves and cultivate a sense of inner peace and tranquility.

Furthermore, frequency techniques can enhance cognitive function, including memory, focus, and creativity. Research has shown that certain frequencies, such as theta waves (4 to 8 Hz), are associated with increased creativity and improved learning abilities. By incorporating frequency-based techniques, such as listening to theta wave music or engaging in brainwave entrainment exercises, we can potentially optimize our cognitive abilities and unlock our full mental potential.

5.1.4 Frequency's Influence on Happiness and Life Satisfaction

Frequency also plays a crucial role in shaping our emotional well-being and overall life satisfaction. By utilizing frequency-based strategies, we can cultivate a positive mindset and enhance our emotional resilience.

For instance, specific frequencies, such as the Solfeggio frequencies, are believed to have a profound impact on our emotional state. These frequencies are said to resonate with the body's energy centers, or chakras, and promote emotional healing and balance. By incorporating Solfeggio frequencies into our daily routine through practices like listening to music or chanting, we can potentially experience a greater sense of happiness, inner peace, and emotional well-being.

Additionally, frequency-based strategies for anxiety relief have gained attention in recent years. Techniques such as deep breathing exercises combined with specific frequencies, like the 528 Hz frequency, have been found to reduce anxiety and promote a state of calmness. By incorporating these techniques into our daily lives, we can effectively manage anxiety and improve our overall emotional well-being.

In conclusion, frequency has a profound impact on various aspects of our lives, including sleep, weight loss, stress, and overall well-being. By understanding and harnessing the power of frequency, we can enhance our mental and physical health, cultivate a positive mindset, and experience a greater sense of overall well-being. Incorporating frequency-based practices into our daily routine can be a transformative journey towards a healthier, happier, and more fulfilling life.

5.2 Frequency's Role in Mental Health

Mental health is a crucial aspect of overall well-being, and the role of frequency in maintaining and improving mental health cannot be overstated. The vibrations and patterns of frequency have a profound impact on our brain and nervous system, influencing our thoughts, emotions, and behaviors. In this section, we will explore the various ways in which frequency affects mental health and discuss the benefits it can bring to individuals seeking to enhance their psychological well-being.

5.2.1 Frequency and Emotional Regulation

One of the key areas where frequency plays a significant role in mental health is emotional regulation. Our emotions are intricately connected to the electrical activity in our brains, and different frequencies can either enhance or disrupt this activity. Research has shown that certain frequencies, such as alpha and theta waves, are associated with a state of relaxation and calmness, promoting emotional stability and reducing anxiety and stress. By incorporating frequency-based techniques into our daily lives, we can regulate our emotions more effectively and experience greater emotional well-being.

5.2.2 Frequency and Mood Enhancement

Another important aspect of mental health is mood regulation. Our mood has a direct impact on our overall outlook on life and our ability to cope with challenges. Frequency can influence our mood by stimulating the release of neurotransmitters such as serotonin and dopamine, which are responsible for feelings of happiness and pleasure. By exposing ourselves to specific frequencies, we can elevate our mood, increase feelings of positivity, and improve our overall mental well-being.

5.2.3 Frequency and Stress Reduction

Stress is a common and often debilitating condition that affects many individuals in today's fast-paced world. The impact of stress on mental health cannot be underestimated, as it can lead to anxiety, depression, and other mental health disorders. Frequency-based techniques have been found to be effective in reducing stress levels by promoting relaxation and activating the body's natural relaxation response. By incorporating frequency practices such as deep breathing exercises, meditation, or listening to calming frequencies, individuals can effectively manage and reduce their stress levels, leading to improved mental health outcomes.

5.2.4 Frequency and Cognitive Function

Cognitive function encompasses various mental processes such as memory, attention, and problem-solving abilities. Research has shown that specific frequencies can enhance cognitive function by improving focus, concentration, and information processing. For example, alpha waves have been associated with increased creativity and improved problem-solving skills, while gamma waves have been linked to heightened mental clarity and improved memory. By utilizing frequency-based techniques, individuals can optimize their cognitive function and enhance their overall mental performance.

5.2.5 Frequency and Mental Disorders

Frequency-based therapies have shown promise in managing and alleviating symptoms of various mental disorders. For instance, studies have demonstrated the effectiveness of specific frequencies in reducing symptoms of depression, anxiety, and post-traumatic stress disorder (PTSD). By targeting the brain's electrical activity through frequency-based interventions, individuals can experience relief from the debilitating symptoms of these disorders and improve their overall mental well-being.

5.2.6 Frequency and Mindfulness

Mindfulness, the practice of being fully present and aware of the present moment, has gained significant attention in recent years for its positive impact on mental health. Frequency can play a role in cultivating mindfulness by helping individuals achieve a state of deep relaxation and heightened awareness. By incorporating frequency-based mindfulness techniques such as binaural beats or guided meditation, individuals can enhance their ability to stay present, reduce rumination, and improve their overall mental well-being.

5.2.7 Frequency and Sleep Disorders

Sleep disorders, such as insomnia and sleep apnea, can have a detrimental effect on mental health. Frequency-based interventions have shown promise in improving sleep quality and reducing the symptoms of these disorders. By utilizing specific frequencies that promote relaxation and induce a state of deep sleep, individuals can experience more restful and rejuvenating sleep, leading to improved mental health outcomes.

In conclusion, frequency plays a significant role in mental health by influencing emotional regulation, mood enhancement, stress reduction, cognitive function, and the management of mental disorders. By incorporating frequency-based techniques into our daily lives, we can optimize our mental well-being and experience greater overall happiness and life satisfaction. The benefits of frequency in mental health are vast, and individuals seeking to enhance their psychological well-being can greatly benefit from exploring and incorporating frequency practices into their daily routines.

5.3 Frequency and Physical Health Benefits

Frequency not only plays a crucial role in sleep, weight loss, and stress management, but it also has a significant impact on physical health. By understanding and harnessing the power of frequency, individuals can unlock a range of benefits that contribute to their overall well-being.

5.3.1 Enhancing Physical Performance and Endurance

One of the key physical health benefits of frequency is its ability to enhance physical performance and endurance. Research has shown that certain frequencies can stimulate the body's energy systems, improving stamina and reducing fatigue during physical activities. By incorporating frequency-based techniques into their training routines, athletes and fitness enthusiasts can optimize their performance and achieve their goals more efficiently.

5.3.2 Managing Pain and Facilitating Rehabilitation

Frequency has also been found to have a positive impact on pain management and rehabilitation. Certain frequencies have analgesic properties, meaning they can alleviate pain and discomfort. This can be particularly beneficial for individuals dealing with chronic pain conditions or recovering from injuries. By utilizing frequency-based approaches, such as targeted sound or vibration therapy, individuals can experience relief from pain and support their rehabilitation process.

5.3.3 Boosting Immune Function

The immune system plays a vital role in protecting the body against infections and diseases. Frequency has been shown to have a direct influence on immune system function, with specific frequencies having immune-boosting effects. By exposing the body to these frequencies, individuals can strengthen their immune response and enhance their overall health. This can be especially beneficial during times of increased vulnerability, such as during the cold and flu season or when recovering from an illness.

5.3.4 Supporting Cellular Regeneration and Healing

Frequency has the potential to support cellular regeneration and healing processes within the body. Certain frequencies have been found to stimulate the production of growth factors and enhance the body's natural healing mechanisms. By incorporating frequency-based techniques, such as low-level laser therapy or electromagnetic field therapy, individuals can promote tissue repair, reduce inflammation, and accelerate the healing process.

5.3.5 Improving Cardiovascular Health

The cardiovascular system is responsible for delivering oxygen and nutrients to the body's tissues and organs. Frequency has been shown to have a positive impact on cardiovascular health by improving blood circulation and reducing the risk of cardiovascular diseases. By utilizing frequency-based techniques, individuals can support their cardiovascular system, maintain healthy blood pressure levels, and reduce the risk of heart-related conditions.

5.3.6 Enhancing Respiratory Function

Frequency can also have a beneficial effect on respiratory function. Certain frequencies have been found to improve lung capacity, support respiratory muscle strength, and enhance overall lung function. By incorporating frequency-based techniques, such as specific breathing exercises or respiratory training devices, individuals can optimize their respiratory health and improve their overall well-being.

5.3.7 Promoting Digestive Health

The digestive system plays a crucial role in nutrient absorption and overall health. Frequency has been shown to have a positive impact on digestive health by promoting optimal digestion, reducing inflammation in the gastrointestinal tract, and supporting the balance of gut bacteria. By incorporating frequency-based techniques, individuals can support their digestive system, alleviate digestive issues, and improve their overall gut health.

5.3.8 Supporting Hormonal Balance

Hormonal balance is essential for overall health and well-being. Frequency has been found to have a regulatory effect on hormone production and balance within the body. By utilizing frequency-based techniques, individuals can support their hormonal health, alleviate symptoms of hormonal imbalances, and promote overall hormonal harmony.

5.3.9 Strengthening Musculoskeletal Health

Frequency can also have a positive impact on musculoskeletal health. Certain frequencies have been shown to promote bone density, support muscle strength and flexibility, and enhance overall musculoskeletal function. By incorporating frequency-based techniques, individuals can support their musculoskeletal system, reduce the risk of injuries, and improve their overall physical performance.

Incorporating frequency-based practices into daily life can have profound effects on physical health. By understanding the various benefits of frequency and utilizing appropriate techniques, individuals can optimize their physical well-being and achieve a higher level of overall health and vitality.

5.4 The Holistic Benefits of Frequency Practices

Frequency practices have a profound impact on various aspects of our well-being, encompassing physical, mental, and emotional health. By understanding and harnessing the power of frequency, individuals can experience a multitude of benefits that contribute to their overall well-being. In this section, we will explore the holistic benefits of frequency practices and how they can enhance our lives.

5.4.1 Improved Sleep Quality

One of the primary benefits of frequency practices is their ability to improve sleep quality. By utilizing specific frequencies, individuals can regulate their brainwaves and promote a state of relaxation conducive to deep and restorative sleep. Research has shown that certain frequencies, such as delta waves, can help individuals fall asleep faster and experience longer periods of uninterrupted sleep. Moreover, frequency-based techniques can also alleviate common sleep disorders, such as insomnia and sleep apnea, leading to a more rejuvenating sleep experience.

5.4.2 Enhanced Weight Loss Efforts

Frequency practices can also have a positive impact on weight loss efforts. By incorporating specific frequencies into daily routines, individuals can stimulate their metabolism, increase fat burning, and optimize their body's ability to shed excess weight. Additionally, frequency-based strategies can help individuals overcome weight loss plateaus by promoting the body's natural ability to break through barriers and continue progressing towards their weight loss goals. By harnessing the power of frequency, individuals can enhance their weight loss journey and achieve sustainable results.

5.4.3 Stress Reduction and Emotional Well-Being

Stress reduction is another significant benefit of frequency practices. Research has shown that certain frequencies can help reduce stress levels by promoting the release of endorphins, which are natural mood-enhancing chemicals in the brain. By incorporating frequency-based stress reduction techniques into daily routines, individuals can experience a greater sense of calmness, relaxation, and emotional well-being. These practices can also help alleviate symptoms of anxiety and depression, providing individuals with a holistic approach to managing their mental health.

5.4.4 Improved Cognitive Function

Frequency practices have been found to have a positive impact on cognitive function. By utilizing specific frequencies, individuals can enhance their focus, concentration, and memory retention. Research suggests that certain frequencies, such as alpha and gamma waves, can improve cognitive performance and information processing. By incorporating frequency-based techniques into their daily lives, individuals can optimize their brain's potential and enhance their overall cognitive abilities.

5.4.5 Boosted Immune System Function

Another significant benefit of frequency practices is their ability to boost immune system function. Research has shown that certain frequencies can stimulate the production of immune cells and enhance the body's natural defense mechanisms. By incorporating frequency-based approaches into their daily routines, individuals can strengthen their immune system, reduce the risk of illness, and promote overall physical health. This is particularly beneficial during times of increased vulnerability, such as during flu seasons or when recovering from an illness.

5.4.6 Enhanced Emotional Well-Being and Life Satisfaction

Frequency practices can also have a profound impact on emotional well-being and life satisfaction. By incorporating frequency-based techniques into daily routines, individuals can cultivate a positive mindset, enhance their emotional resilience, and experience greater overall life satisfaction. These practices can help individuals manage stress, regulate their emotions, and foster a sense of inner peace and contentment. By aligning their frequencies with positive emotions, individuals can create a harmonious and fulfilling life experience.

5.4.7 Overall Well-Being and Quality of Life

The holistic benefits of frequency practices ultimately contribute to an overall sense of well-being and improved quality of life. By optimizing sleep patterns, enhancing weight loss efforts, reducing stress, improving cognitive function, boosting immune system function, and promoting emotional well-being, individuals can experience a comprehensive transformation in their lives. Frequency practices provide a holistic approach to well-being, addressing various aspects of health and empowering individuals to live their lives to the fullest.

Incorporating frequency practices into daily routines can lead to a profound shift in overall well-being, allowing individuals to unlock their full potential and experience a greater sense of vitality, happiness, and fulfillment. By understanding and harnessing the power of frequency, individuals can embark on a transformative journey towards optimal health and well-being.

6 Frequency and Well-Being

6.1 The Interplay Between Frequency and Well-Being

Frequency, in its various forms, has a profound impact on our overall well-being. It influences our physical health, mental state, and emotional balance. In this section, we will explore the interplay between frequency and well-being, delving into its influence on happiness, life satisfaction, and the cultivation of a positive mindset. We will also discuss frequency-based strategies for enhancing overall well-being.

6.1.1 Frequency's Influence on Happiness and Life Satisfaction

Happiness and life satisfaction are fundamental aspects of well-being. Research has shown that frequency plays a significant role in shaping our emotional state and overall happiness levels. The vibrational energy emitted by different frequencies can directly affect our mood, emotions, and overall outlook on life.

When we are exposed to high-frequency vibrations, such as those found in positive thoughts, uplifting music, or nature sounds, our brain releases neurotransmitters like serotonin and dopamine, which are associated with feelings of happiness and well-being. These frequencies have the power to uplift our spirits, increase our energy levels, and promote a positive mindset.

Conversely, low-frequency vibrations, such as negative thoughts, stressful environments, or dissonant sounds, can have a detrimental effect on our emotional well-being. They can lead to feelings of sadness, anxiety, and even depression. By understanding the impact of frequency on our emotional state, we can consciously choose to surround ourselves with positive vibrations and create a more joyful and fulfilling life.

6.1.2 Using Frequency to Cultivate a Positive Mindset

Cultivating a positive mindset is essential for overall well-being. Our thoughts and beliefs shape our reality, and frequency can play a pivotal role in influencing our mindset. By consciously aligning ourselves with high-frequency vibrations, we can shift our perspective, enhance our resilience, and cultivate a positive outlook on life.

One effective way to harness the power of frequency for a positive mindset is through affirmations. Affirmations are positive statements that we repeat to ourselves, reinforcing empowering beliefs and intentions. By infusing these affirmations with high-frequency vibrations, we can amplify their impact and create a powerful shift in our mindset.

Another powerful technique is visualization. By visualizing positive outcomes and experiences while immersing ourselves in high-frequency vibrations, we can enhance the manifestation of our desires and create a positive mental state. Visualization combined with frequency-based practices, such as meditation or listening to uplifting music, can help rewire our brain and create new neural pathways that support a positive mindset.

6.1.3 Frequency-Based Strategies for Enhancing Overall Well-Being

In addition to influencing happiness and mindset, frequency has a wide range of effects on our overall well-being. By incorporating frequency-based strategies into our daily lives, we can enhance our physical health, mental clarity, and emotional balance.

One such strategy is sound therapy. Sound therapy utilizes specific frequencies and vibrations to promote relaxation, reduce stress, and restore balance to the body and mind. Techniques such as binaural beats, where two slightly different frequencies are played simultaneously to create a desired brainwave state, can help induce deep relaxation, improve sleep quality, and enhance overall well-being.

Another effective strategy is mindfulness meditation. By focusing our attention on the present moment and immersing ourselves in the frequency of the present, we can cultivate a state of calmness, reduce stress, and improve overall well-being. Mindfulness practices, combined with frequency-based techniques such as deep breathing exercises or guided meditations, can have a profound impact on our mental and emotional state.

Additionally, incorporating frequency-based practices into our physical exercise routine can enhance our overall well-being. For example, engaging in activities such as yoga or tai chi, which emphasize the flow of energy and the synchronization of breath with movement, can help balance our energy centers and promote a sense of well-being. These practices, combined with the intentional use of high-frequency music or nature sounds, can create a harmonious and uplifting experience.

In conclusion, frequency has a significant impact on our well-being. It influences our happiness, life satisfaction, and overall mindset. By consciously aligning ourselves with high-frequency vibrations and incorporating frequency-based strategies into our daily lives, we can enhance our overall well-being and create a more fulfilling and joyful existence.

6.2 Frequency's Influence on Happiness and Life Satisfaction

Frequency, the measure of how often something occurs, plays a significant role in shaping our happiness and overall life satisfaction. The vibrations and patterns of frequency have a profound impact on our mental and emotional well-being, influencing our mood, mindset, and overall outlook on life. In this section, we will explore the ways in which frequency can influence happiness and life satisfaction, and how we can harness its power to cultivate a positive mindset and enhance our overall well-being.

The Connection Between Frequency and Emotional State

Our emotions are intricately linked to the frequencies we experience. Different frequencies can evoke various emotional responses, ranging from joy and contentment to sadness and anxiety. When we are exposed to high-frequency vibrations, such as uplifting music or positive interactions, our emotional state tends to be more positive and optimistic. Conversely, low-frequency vibrations, such as negative thoughts or stressful situations, can lead to feelings of sadness, frustration, or anxiety.

The Role of Frequency in Shaping Mindset

Our mindset, or the way we perceive and interpret the world around us, is heavily influenced by the frequencies we encounter. Positive frequencies can help us develop a growth mindset, characterized by resilience, optimism, and a belief in our ability to overcome challenges. On the other hand, negative frequencies can contribute to a fixed mindset, where we feel stuck, helpless, and unable to change our circumstances.

By consciously exposing ourselves to positive frequencies, such as uplifting music, inspiring literature, or supportive social connections, we can cultivate a positive mindset that fosters happiness and life satisfaction. These positive frequencies can help us reframe challenges as opportunities for growth, enhance our self-belief, and increase our resilience in the face of adversity.

Frequency's Impact on Neural Pathways

The frequencies we experience can also influence the neural pathways in our brain, shaping our thoughts, beliefs, and behaviors. When we consistently expose ourselves to positive frequencies, we strengthen the neural pathways associated with happiness, gratitude, and well-being. This can lead to a more positive and optimistic outlook on life, as well as an increased capacity for experiencing joy and contentment.

Conversely, exposure to negative frequencies can reinforce neural pathways associated with stress, anxiety, and negativity. This can create a cycle of negative thinking and emotional distress, impacting our overall happiness and life satisfaction. By consciously choosing to engage with positive frequencies, we can rewire our neural pathways and create a more positive and fulfilling life experience.

Harnessing the Power of Frequency for Happiness and Life Satisfaction

To harness the power of frequency for happiness and life satisfaction, it is essential to be intentional about the frequencies we expose ourselves to on a daily basis. Here are some strategies to incorporate frequency into our lives:

1. Surround Yourself with Positive Frequencies

Surround yourself with positive people, uplifting music, inspiring books, and motivational content. Seek out activities and environments that resonate with positive frequencies, such as spending time in nature, practicing gratitude, and engaging in creative pursuits. By consciously choosing positive frequencies, we can create an environment that supports our happiness and well-being.

2. Practice Mindfulness and Self-Awareness

Developing mindfulness and self-awareness allows us to observe our thoughts, emotions, and the frequencies we are experiencing. By cultivating present moment awareness, we can identify negative frequencies and consciously choose to shift our focus to more positive ones. Mindfulness practices such as meditation, deep breathing, and body scans can help us become more attuned to the frequencies we are experiencing and make intentional choices to cultivate happiness and life satisfaction.

3. Cultivate Positive Relationships

Surround yourself with individuals who radiate positive frequencies and support your well-being. Seek out relationships that uplift and inspire you, and distance yourself from toxic or negative influences. Positive social connections can have a profound impact on our happiness and life satisfaction, as they provide a supportive network and opportunities for growth and connection.

4. Engage in Frequency-Based Practices

Explore frequency-based practices such as sound therapy, binaural beats, or guided visualizations. These practices utilize specific frequencies to induce relaxation, reduce stress, and promote a positive mindset. Incorporating these practices into your daily routine can help you align with positive frequencies and enhance your overall well-being.

5. Practice Gratitude and Positive Affirmations

Cultivating gratitude and positive affirmations can help shift our focus towards positive frequencies. Take time each day to reflect on the things you are grateful for and affirm positive beliefs about yourself and your life. By consciously directing your attention towards positive frequencies, you can rewire your brain and cultivate a more positive and fulfilling life experience.

In conclusion, frequency has a profound influence on our happiness and life satisfaction. By understanding the connection between frequency and our emotional state, mindset, and neural pathways, we can harness its power to cultivate a positive mindset and enhance our overall well-being. By intentionally choosing positive frequencies, practicing mindfulness, cultivating positive relationships, engaging in frequency-based practices, and practicing gratitude, we can align ourselves with the frequencies that promote happiness and life satisfaction.

6.3 Using Frequency to Cultivate a Positive Mindset

Frequency not only has a profound impact on our physical health but also plays a crucial role in shaping our mental and emotional well-being. By understanding and harnessing the power of frequency, we can cultivate a positive mindset that promotes happiness, resilience, and overall well-being. In this section, we will explore how frequency can be used to enhance our mental state and provide practical strategies for incorporating frequency practices into our daily lives.

6.3.1 The Influence of Frequency on Mental State

The frequency at which our brain operates has a direct influence on our mental state. Different frequencies correspond to different states of consciousness, ranging from deep sleep to heightened alertness. By consciously manipulating these frequencies, we can positively impact our mood, cognitive function, and emotional well-being.

6.3.2 Harnessing Frequency for Positive Emotions

One of the most powerful ways to use frequency to cultivate a positive mindset is through the practice of binaural beats. Binaural beats are created by playing two slightly different frequencies in each ear, which then interact in the brain to produce a third frequency. This third frequency corresponds to a specific brainwave state, such as alpha or theta, which is associated with relaxation, creativity, and positive emotions.

Listening to binaural beats can help reduce stress, anxiety, and negative thought patterns, while promoting feelings of calmness, joy, and overall well-being. Incorporating binaural beats into our daily routine, whether through dedicated meditation sessions or simply as background music during daily activities, can have a profound impact on our mental state.

6.3.3 Frequency-Based Mindfulness Practices

Mindfulness, the practice of being fully present in the moment, is another powerful tool for cultivating a positive mindset. By focusing our attention on the present moment, we can reduce stress, increase self-awareness, and enhance our overall well-being. Frequency can be used as a support for mindfulness practices, helping to deepen our state of presence and awareness.

One effective technique is to use frequency-based guided meditations or visualizations. These guided practices often incorporate soothing sounds or music that resonate at specific frequencies, helping to induce a state of relaxation and focus. By immersing ourselves in these frequency-enhanced mindfulness practices, we can cultivate a positive mindset and develop a greater sense of inner peace and clarity.

6.3.4 Frequency and Affirmations

Affirmations are positive statements that we repeat to ourselves to reinforce desired beliefs or behaviors. When combined with the power of frequency, affirmations can become even more potent in shaping our mindset. By using frequency-enhanced affirmations, we can tap into the subconscious mind and reprogram negative thought patterns, replacing them with positive and empowering beliefs.

Listening to affirmations recorded at specific frequencies, such as theta or gamma, can help bypass the conscious mind's resistance and directly influence the subconscious. This can lead to a profound shift in our mindset, promoting self-confidence, motivation, and a positive outlook on life.

6.3.5 Incorporating Frequency Practices into Daily Life

To fully harness the benefits of frequency for cultivating a positive mindset, it is essential to integrate frequency practices into our daily lives. Here are some practical strategies for incorporating frequency into our routines:

1. Morning Ritual: Start your day with a frequency-based meditation or affirmation practice. Set the tone for the day by aligning your mindset with positivity and intention.
2. Mindful Breaks: Take short breaks throughout the day to engage in frequency-enhanced mindfulness practices. This can be as simple as closing your eyes, taking deep breaths, and focusing on the present moment.
3. Evening Wind-Down: Create a relaxing evening routine that includes frequency-based practices such as listening to calming music or guided meditations. This can help promote a restful sleep and prepare your mind for a positive start to the next day.
4. Consistency is Key: Incorporate frequency practices into your daily routine consistently. Just like any habit, the more you practice, the more profound the effects will be on your mindset and overall well-being.

By consciously incorporating frequency practices into our lives, we can cultivate a positive mindset that supports our mental and emotional well-being. Whether through binaural beats, mindfulness practices, or frequency-enhanced affirmations, the power of frequency can be harnessed to promote happiness, resilience, and a greater sense of overall well-being.

6.4 Frequency-Based Strategies for Enhancing Overall Well-Being

Frequency plays a significant role in our overall well-being, influencing various aspects of our physical, mental, and emotional health. By understanding the impact of frequency on different areas of our lives, we can develop effective strategies to enhance our overall well-being. In this section, we will explore frequency-based strategies that can help improve our well-being and lead to a more fulfilling and balanced life.

6.4.1 Cultivating Positive Mindset through Frequency

One of the key strategies for enhancing overall well-being is to cultivate a positive mindset. Our thoughts and beliefs have a profound impact on our emotions, behaviors, and overall outlook on life. By harnessing the power of frequency, we can shift our mindset towards positivity and create a more optimistic and resilient approach to life.

Frequency-based techniques such as affirmations and positive self-talk can help rewire our thought patterns and replace negative thoughts with empowering and uplifting ones. By consistently exposing ourselves to positive frequencies through affirmations or listening to uplifting music, we can reprogram our subconscious mind and cultivate a more positive mindset.

6.4.2 Enhancing Emotional Well-Being with Frequency

Emotional well-being is crucial for overall well-being, and frequency can play a significant role in regulating our emotions. Certain frequencies have been found to have a calming and soothing effect on the mind and body, helping to reduce stress, anxiety, and negative emotions.

Practicing frequency-based techniques such as sound therapy or binaural beats can help induce a state of relaxation and promote emotional balance. These techniques involve listening to specific frequencies that resonate with different emotions, helping to release emotional blockages and promote a sense of calm and well-being.

6.4.3 Promoting Physical Health through Frequency

Frequency not only impacts our mental and emotional well-being but also has a profound effect on our physical health. By harnessing the power of frequency, we can enhance our physical well-being and support our body's natural healing processes.

Certain frequencies have been found to stimulate the body's natural healing mechanisms and promote physical health. For example, frequency-based therapies such as electromagnetic field therapy or pulsed electromagnetic field therapy (PEMF) have shown promising results in promoting tissue regeneration, reducing inflammation, and accelerating the healing process.

6.4.4 Strengthening Relationships with Frequency

Healthy and fulfilling relationships are essential for overall well-being. Frequency-based strategies can also be applied to enhance our relationships and create deeper connections with others.

By practicing active listening and being present in our interactions, we can attune ourselves to the frequency of the person we are communicating with, fostering a deeper understanding and connection. Additionally, engaging in activities that promote positive frequencies, such as shared hobbies or engaging in acts of kindness, can strengthen the bond between individuals and contribute to overall well-being.

6.4.5 Nurturing Spiritual Well-Being through Frequency

Spiritual well-being is an integral part of overall well-being, and frequency can be a powerful tool for nurturing our spiritual growth. By aligning ourselves with higher frequencies, we can deepen our connection to our inner selves and the world around us.

Practices such as meditation, chanting, or listening to sacred music can help raise our vibrational frequency and facilitate spiritual growth. These practices create a space for introspection, self-reflection, and connection to something greater than ourselves, leading to a sense of purpose, meaning, and fulfillment.

6.4.6 Creating a Frequency-Based Self-Care Routine

Incorporating frequency-based practices into our daily self-care routine can significantly contribute to our overall well-being. By dedicating time each day to engage in activities that promote positive frequencies, we can prioritize our well-being and create a foundation for a balanced and fulfilling life.

Some frequency-based self-care practices include journaling, practicing gratitude, engaging in mindfulness or breathwork exercises, and spending time in nature. These activities help us reconnect with ourselves, reduce stress, and promote a sense of inner peace and well-being.

6.4.7 Seeking Professional Guidance

While frequency-based strategies can be beneficial for enhancing overall well-being, it is important to seek professional guidance when necessary. If you are experiencing significant challenges in any area of your well-being, such as sleep, weight loss, or stress management, consulting with healthcare professionals or experts in the field can provide personalized guidance and support.

Remember, everyone's journey to well-being is unique, and it may take time to find the frequency-based strategies that work best for you. Be patient, stay open-minded, and embrace the power of frequency as a tool for enhancing your overall well-being.

In the next chapter, we will delve into frequency and sleep optimization techniques, exploring how frequency can be utilized to improve sleep duration, quality, and overcome sleep disorders.

7 Frequency and Sleep Optimization Techniques

7.1 Understanding Sleep Optimization through Frequency

Sleep is a fundamental aspect of our overall well-being, and its importance cannot be overstated. It is during sleep that our bodies repair and rejuvenate, and our minds process and consolidate information. However, in today's fast-paced and demanding world, many individuals struggle with achieving restful and restorative sleep. This is where the concept of frequency comes into play.

Frequency, in the context of sleep optimization, refers to the specific patterns and rhythms of brainwave activity that occur during different stages of sleep. These brainwave frequencies are categorized into four main types: delta, theta, alpha, and beta. Each frequency range corresponds to a different state of consciousness and has a unique impact on our sleep quality.

7.1.1 The Role of Brainwave Frequencies in Sleep

Delta waves are the slowest brainwave frequencies and are associated with deep sleep. When we experience deep sleep, our bodies undergo essential physiological processes such as tissue repair, hormone regulation, and immune system strengthening. Delta waves are crucial for promoting physical restoration and overall well-being.

Theta waves are slightly faster than delta waves and are associated with the dream-like state of REM (rapid eye movement) sleep. During REM sleep, our brains are highly active, and this stage is essential for cognitive processes such as memory consolidation and emotional regulation. Theta waves play a vital role in facilitating these cognitive functions during REM sleep.

Alpha waves are even faster than theta waves and are associated with relaxed wakefulness and light sleep. When we are in a state of relaxation, such as during meditation or before falling asleep, alpha waves dominate our brainwave activity. Alpha waves promote a calm and peaceful state of mind, making it easier to transition into sleep.

Beta waves are the fastest brainwave frequencies and are associated with wakefulness and alertness. When we are awake and engaged in focused mental activity, beta waves dominate our brainwave activity. However, excessive beta wave activity during sleep can lead to restlessness and difficulty falling asleep.

7.1.2 The Impact of Frequency on Sleep Quality

The balance and synchronization of these brainwave frequencies are crucial for achieving optimal sleep quality. When our brainwave frequencies are in harmony, we experience deep, restful sleep, and wake up feeling refreshed and rejuvenated. However, imbalances or disruptions in these frequencies can lead to various sleep disturbances, such as insomnia, sleep apnea, or restless leg syndrome.

For example, individuals with an overabundance of beta wave activity during sleep may struggle with racing thoughts, anxiety, and difficulty falling asleep. On the other hand, those with insufficient delta wave activity may experience shallow sleep, frequent awakenings, and a lack of restorative rest.

7.1.3 Using Frequency to Improve Sleep Duration and Quality

Understanding the impact of frequency on sleep allows us to harness its power to optimize our sleep duration and quality. By consciously influencing our brainwave frequencies, we can promote deep sleep, enhance cognitive processes during REM sleep, and create a calm and relaxed state conducive to falling asleep.

One effective technique for improving sleep through frequency is binaural beats. Binaural beats involve listening to two slightly different frequencies in each ear, which creates a third frequency in the brain. This third frequency corresponds to the desired brainwave state, such as delta or theta waves, promoting relaxation and sleep.

Another technique is the use of specific sound frequencies, such as white noise or nature sounds, to create a soothing and calming environment for sleep. These frequencies can help drown out external disturbances and promote a sense of tranquility, making it easier to fall asleep and stay asleep throughout the night.

7.1.4 Frequency Techniques for Overcoming Sleep Disorders

For individuals struggling with sleep disorders, frequency-based techniques can offer a natural and non-invasive approach to overcoming these challenges. For example, individuals with insomnia may benefit from listening to delta wave binaural beats or using white noise machines to promote deep sleep.

Those with sleep apnea or restless leg syndrome may find relief through frequency-based relaxation techniques, such as guided meditation or progressive muscle relaxation. These techniques can help reduce muscle tension, promote relaxation, and improve overall sleep quality.

In conclusion, understanding the role of frequency in sleep optimization is essential for achieving restful and rejuvenating sleep. By harnessing the power of brainwave frequencies, we can enhance sleep duration, improve sleep quality, and overcome sleep disorders. Incorporating frequency-based techniques into our sleep routines can lead to profound improvements in our overall well-being and quality of life.

7.2 Frequency-Based Sleep Hygiene Practices

Sleep is a fundamental aspect of our overall well-being, and the quality and duration of our sleep can have a significant impact on our physical and mental health. In recent years, researchers have begun to explore the role of frequency in optimizing sleep patterns and improving sleep quality. By understanding and harnessing the power of frequency, we can develop effective sleep hygiene practices that promote restful and rejuvenating sleep.

7.2.1 Creating a Sleep-Friendly Environment

One of the key aspects of frequency-based sleep hygiene practices is creating a sleep-friendly environment. This involves ensuring that your bedroom is a calm and relaxing space that promotes sleep. Consider the following tips:

- **Eliminate electronic devices**: Electronic devices emit blue light, which can disrupt the production of melatonin, a hormone that regulates sleep. Remove electronic devices such as smartphones, tablets, and laptops from your bedroom to create a technology-free zone.
- **Optimize room temperature**: The temperature of your bedroom can significantly impact your sleep quality. Keep your bedroom cool, ideally between 60 to 67 degrees Fahrenheit (15 to 19 degrees Celsius), to promote better sleep.
- **Reduce noise**: Excessive noise can disrupt your sleep. Use earplugs or a white noise machine to block out unwanted sounds and create a peaceful sleep environment.
- **Ensure darkness**: Darkness is essential for the production of melatonin. Use blackout curtains or an eye mask to block out any external light sources that may interfere with your sleep.

7.2.2 Establishing a Consistent Sleep Routine

Consistency is key when it comes to optimizing sleep patterns. By establishing a regular sleep routine, you can train your body to recognize when it's time to sleep and wake up. Consider the following tips:

- **Set a consistent sleep schedule**: Go to bed and wake up at the same time every day, even on weekends. This helps regulate your body's internal clock and promotes better sleep quality.
- **Create a pre-sleep routine**: Develop a relaxing pre-sleep routine that signals to your body that it's time to wind down. This may include activities such as reading a book, taking a warm bath, or practicing relaxation techniques.
- **Avoid stimulating activities before bed**: Engaging in stimulating activities, such as intense exercise or watching thrilling movies, close to bedtime can make it difficult to fall asleep. Instead, opt for calming activities that promote relaxation.

7.2.3 Managing Stress and Anxiety

Stress and anxiety can significantly impact sleep quality. Frequency-based sleep hygiene practices can help manage stress and promote a more peaceful state of mind before bed. Consider the following techniques:

- **Breathing exercises**: Deep breathing exercises, such as diaphragmatic breathing or the 4-7-8 technique, can help activate the body's relaxation response and reduce stress and anxiety.
- **Meditation and mindfulness**: Practicing meditation or mindfulness techniques before bed can help calm the mind and promote a sense of tranquility. Focus on your breath or engage in guided meditation to quiet racing thoughts.
- **Progressive muscle relaxation**: Progressive muscle relaxation involves tensing and then releasing each muscle group in your body, promoting physical and mental relaxation. This technique can be

particularly helpful for individuals who experience muscle tension or restlessness before bed.

7.2.4 Optimizing Sleep Environment

In addition to creating a sleep-friendly environment, there are other frequency-based practices that can optimize your sleep environment. Consider the following techniques:

- **Aromatherapy**: Certain scents, such as lavender, chamomile, or bergamot, have been shown to promote relaxation and improve sleep quality. Use essential oils or a diffuser to infuse your bedroom with these calming scents.
- **Sleep-inducing sounds**: White noise, nature sounds, or soothing music can help drown out external noises and create a peaceful sleep environment. Experiment with different sounds to find what works best for you.
- **Comfortable bedding**: Investing in comfortable bedding, including a supportive mattress and pillows, can significantly improve sleep quality. Choose materials that are breathable and promote temperature regulation for optimal comfort.

By incorporating these frequency-based sleep hygiene practices into your daily routine, you can enhance the quality and duration of your sleep. Remember, consistency is key, so be patient and allow your body to adjust to these new habits. With time, you will experience the positive effects of frequency on your sleep and overall well-being.

7.3 Using Frequency to Improve Sleep Duration and Quality

Sleep is a fundamental aspect of our overall well-being, and its importance cannot be overstated. It is during sleep that our bodies repair and rejuvenate, and our minds process and consolidate information. However, many individuals struggle with sleep-related issues, such as difficulty falling asleep, staying asleep, or experiencing restful sleep. This is where the power of frequency comes into play. By harnessing the right frequencies, we can improve both the duration and quality of our sleep.

7.3.1 Understanding the Role of Frequency in Sleep

Frequency, in the context of sleep, refers to the specific vibrations or oscillations that occur in our brain and body during different stages of sleep. These frequencies can be measured using electroencephalography (EEG) and are categorized into different brainwave patterns, including delta, theta, alpha, beta, and gamma waves. Each of these brainwave patterns corresponds to a different state of consciousness and plays a crucial role in our sleep architecture.

7.3.2 The Impact of Frequency on Sleep Duration

One of the ways frequency affects sleep is by influencing the duration of our sleep cycles. Sleep cycles typically last around 90 minutes and consist of different stages, including light sleep, deep sleep, and rapid eye movement (REM) sleep. Each stage serves a unique purpose in restoring and maintaining our physical and mental well-being.

By using specific frequencies, we can optimize the duration of each sleep stage, ensuring that we spend adequate time in deep sleep and REM sleep.

Deep sleep is essential for physical restoration, immune function, and memory consolidation, while REM sleep is crucial for cognitive processing, emotional regulation, and dreaming. By enhancing the duration of these vital sleep stages, we can wake up feeling more refreshed and rejuvenated.

7.3.3 Enhancing Sleep Quality with Frequency

In addition to influencing sleep duration, frequency also plays a significant role in improving the quality of our sleep. The quality of sleep refers to how restful and restorative our sleep is, and it is influenced by factors such as sleep disturbances, sleep fragmentation, and the presence of sleep disorders.

Certain frequencies, such as those in the delta and theta range, have a calming and soothing effect on the brain and body. By listening to or experiencing these frequencies through techniques like binaural beats or isochronic tones, we can induce a state of relaxation and promote a more peaceful sleep environment. These frequencies help to reduce anxiety, quiet the mind, and alleviate stress, allowing us to fall asleep more easily and experience deeper, more restorative sleep.

7.3.4 Frequency-Based Techniques for Improving Sleep

There are several frequency-based techniques that can be incorporated into our bedtime routine to improve sleep duration and quality:

7.3.4.1 Binaural Beats and Isochronic Tones

Binaural beats and isochronic tones are two popular techniques that utilize frequency to influence brainwave patterns and induce specific states of consciousness. Binaural beats involve listening to two slightly different frequencies in each ear, which the brain then perceives as a single beat. This beat corresponds to a specific brainwave frequency and can help guide the brain into a desired state, such as deep relaxation or sleep.

Isochronic tones, on the other hand, involve listening to evenly spaced pulses of sound or light, which also correspond to specific brainwave frequencies. These tones are often more effective than binaural beats for inducing relaxation and sleep.

7.3.4.2 White Noise and Nature Sounds

White noise and nature sounds can also be used to improve sleep quality. These sounds create a consistent background noise that helps mask other disruptive sounds and promotes a more peaceful sleep environment. White noise machines or apps that generate sounds like rainfall, ocean waves, or gentle forest sounds can be particularly helpful for individuals who are easily disturbed by external noises.

7.3.4.3 Meditation and Mindfulness Practices

Meditation and mindfulness practices can also be beneficial for improving sleep. These practices involve focusing the mind on the present moment and cultivating a sense of calm and relaxation. By incorporating frequency-based meditation techniques, such as chanting or using specific mantras, we can further enhance the relaxation response and promote better sleep.

7.3.5 Frequency Techniques for Overcoming Sleep Disorders

For individuals who suffer from sleep disorders, such as insomnia or sleep apnea, frequency-based techniques can offer relief and support. By targeting the specific brainwave patterns associated with these disorders, frequency can help regulate sleep cycles and promote healthier sleep patterns.

For example, individuals with insomnia may benefit from listening to frequencies in the alpha or theta range before bed to induce a state of relaxation and ease the transition into sleep. Those with sleep apnea may find relief through techniques that promote deep, restorative sleep, such as delta wave stimulation.

In conclusion, frequency plays a significant role in improving both the duration and quality of our sleep. By incorporating frequency-based techniques, such as binaural beats, white noise, and meditation, we can enhance our sleep experience and wake up feeling more refreshed and rejuvenated. Whether we are looking to optimize our sleep duration, overcome sleep disorders, or simply improve our overall well-being, harnessing the power of frequency can be a valuable tool in our sleep optimization toolkit.

7.4 Frequency Techniques for Overcoming Sleep Disorders

Sleep disorders can have a significant impact on our overall well-being and quality of life. They can disrupt our sleep patterns, leaving us feeling tired, irritable, and unable to function at our best. Fortunately, frequency techniques can be used to overcome sleep disorders and promote restful, rejuvenating sleep. In this section, we will explore some effective frequency techniques that can help you overcome sleep disorders and improve your sleep quality.

7.4.1 Binaural Beats

Binaural beats are a popular frequency technique used to promote relaxation and improve sleep. They involve listening to two slightly different frequencies in each ear, which creates a third frequency in the brain. This third frequency corresponds to a specific brainwave state, such as alpha or theta, which are associated with relaxation and deep sleep. By listening to binaural beats before bed, you can help synchronize your brainwaves and induce a state of deep relaxation, making it easier to fall asleep and stay asleep throughout the night.

7.4.2 Isochronic Tones

Similar to binaural beats, isochronic tones are another frequency technique that can help overcome sleep disorders. Unlike binaural beats, isochronic tones use a single tone that is turned on and off at regular intervals. This rhythmic pattern helps to entrain the brainwaves and promote a state of relaxation and sleep. By listening to isochronic tones before bed, you can help calm your mind, reduce racing thoughts, and prepare your body for a restful night's sleep.

7.4.3 White Noise

White noise is a frequency technique that involves playing a consistent, soothing sound in the background while you sleep. This sound can help mask other noises that may be disrupting your sleep, such as traffic or snoring. White noise can also help create a calming environment that promotes relaxation and sleep. There are various white noise machines and apps available that offer a range of sounds, from gentle rain to ocean waves, allowing you to find the one that works best for you.

7.4.4 Guided Meditation

Guided meditation is a frequency technique that combines soothing voice guidance with calming background music or sounds. It can help quiet the mind, reduce stress, and promote a state of relaxation conducive to sleep. By following the instructions of a guided meditation, you can focus your attention on your breath, body sensations, or visualizations, helping to calm your mind and prepare your body for sleep. There are numerous guided meditation apps and recordings available that specifically target sleep and relaxation.

7.4.5 Progressive Muscle Relaxation

Progressive muscle relaxation is a technique that involves systematically tensing and relaxing different muscle groups in the body. By consciously tensing and then releasing the tension in each muscle group, you can promote physical and mental relaxation, making it easier to fall asleep. This technique can help release tension and reduce muscle stiffness, allowing your body to enter a state of deep relaxation. There are guided progressive muscle relaxation recordings and apps available that can guide you through the process.

7.4.6 Acupuncture

Acupuncture is an ancient Chinese practice that involves inserting thin needles into specific points on the body. It is believed to help balance the flow of energy, or qi, in the body, promoting overall well-being and health. Acupuncture has been found to be effective in treating various sleep disorders, such as insomnia and sleep apnea. By stimulating specific acupuncture points, the body's natural sleep mechanisms can be activated, helping to regulate sleep patterns and improve sleep quality.

7.4.7 Aromatherapy

Aromatherapy involves using essential oils derived from plants to promote relaxation and improve sleep. Certain essential oils, such as lavender, chamomile, and bergamot, have calming properties that can help reduce anxiety, promote relaxation, and improve sleep quality. You can use essential oils in a diffuser, add a few drops to your bath, or apply them topically (diluted with a carrier oil) before bed. Aromatherapy can create a soothing environment that signals to your body and mind that it's time to unwind and prepare for sleep.

7.4.8 Sleep Hygiene Practices

In addition to frequency techniques, it's important to incorporate good sleep hygiene practices into your routine to overcome sleep disorders. These practices include maintaining a consistent sleep schedule, creating a comfortable sleep environment, avoiding stimulating activities before bed, limiting exposure to electronic devices, and practicing relaxation techniques. By combining frequency techniques with good sleep hygiene practices, you can create an optimal sleep environment and improve your sleep quality.

Incorporating these frequency techniques into your daily routine can help you overcome sleep disorders and achieve restful, rejuvenating sleep. Experiment with different techniques and find what works best for you. Remember, consistency is key, so make these techniques a regular part of your bedtime routine. By prioritizing your sleep and taking steps to improve it, you can enhance your overall well-being and enjoy the benefits of a good night's rest.

8 Frequency and Weight Loss Strategies

8.1 Incorporating Frequency into Weight Loss Plans

Weight loss is a common goal for many individuals, and incorporating frequency into weight loss plans can be a powerful strategy to enhance results. Frequency, in this context, refers to the specific vibrations or oscillations that can have an impact on various physiological processes in the body. By understanding and harnessing the power of frequency, individuals can optimize their weight loss efforts and achieve sustainable results.

8.1.1 Understanding the Role of Frequency in Weight Loss

Frequency plays a crucial role in weight loss by influencing several key factors that contribute to the body's ability to burn fat and maintain a healthy weight. One of the primary ways frequency affects weight loss is through its impact on metabolism. Metabolism refers to the chemical processes that occur within the body to convert food into energy. By increasing the body's metabolic rate, frequency can help individuals burn calories more efficiently, leading to weight loss.

8.1.2 Frequency-Based Dietary Approaches for Weight Management

Incorporating frequency into weight loss plans involves not only understanding the role of frequency but also adopting specific dietary approaches that align with its principles. One such approach is mindful eating, which involves paying close attention to the frequency and quality of food consumed. By being mindful of the frequency at which meals and snacks are consumed, individuals can regulate their hunger and satiety cues, leading to better portion control and reduced calorie intake.

Another frequency-based dietary approach is focusing on nutrient-dense foods. These are foods that are rich in essential nutrients while being relatively low in calories. By incorporating a variety of fruits, vegetables, whole grains, lean proteins, and healthy fats into their diet, individuals can ensure they are providing their body with the necessary nutrients for optimal functioning while keeping their calorie intake in check.

8.1.3 Using Frequency to Boost Metabolism and Fat Burning

Frequency can be utilized to boost metabolism and enhance the body's ability to burn fat. One effective strategy is incorporating high-intensity interval training (HIIT) into a weight loss plan. HIIT involves alternating between short bursts of intense exercise and periods of rest or lower intensity. This type of exercise has been shown to increase the body's metabolic rate and promote fat burning even after the workout is complete.

Additionally, incorporating strength training exercises into a weight loss plan can help increase muscle mass. Muscles are more metabolically active than fat, meaning they burn more calories at rest. By building lean muscle through strength training, individuals can increase their basal metabolic rate, leading to more efficient calorie burning throughout the day.

8.1.4 Frequency Techniques for Overcoming Weight Loss Plateaus

Weight loss plateaus are common and can be frustrating for individuals on a weight loss journey. However, by incorporating frequency techniques, individuals can overcome these plateaus and continue making progress towards their goals. One such technique is varying the intensity and duration of workouts. By challenging the body with different types of exercises and varying the intensity, individuals can prevent their metabolism from adapting to a specific routine, thus promoting continued weight loss.

Another technique is incorporating intermittent fasting into a weight loss plan. Intermittent fasting involves cycling between periods of fasting and eating. This approach has been shown to have various benefits, including improved insulin sensitivity, increased fat burning, and reduced calorie intake. By incorporating intermittent fasting, individuals can break through weight loss plateaus and continue making progress towards their goals.

Incorporating frequency into weight loss plans can be a powerful strategy for optimizing results. By understanding the role of frequency in weight loss, adopting frequency-based dietary approaches, utilizing frequency to boost metabolism and fat burning, and employing frequency techniques to overcome plateaus, individuals can enhance their weight loss journey and achieve long-term success. It is important to consult with a healthcare professional or a registered dietitian before making any significant changes to your weight loss plan.

8.2 Frequency-Based Dietary Approaches for Weight Management

When it comes to weight management, many people focus solely on calorie intake and exercise. However, emerging research suggests that the frequency of our meals and the timing of our eating patterns can also play a significant role in weight loss and maintenance. In this section, we will explore the concept of frequency-based dietary approaches and how they can be utilized to support weight management goals.

8.2.1 The Importance of Meal Frequency

Traditionally, the three-meals-a-day approach has been the norm for many individuals. However, recent studies have shown that increasing the frequency of meals throughout the day can have a positive impact on weight management. By consuming smaller, more frequent meals, individuals can help regulate their blood sugar levels, prevent excessive hunger, and maintain a steady metabolism.

8.2.2 The Role of Macronutrients

In addition to meal frequency, the composition of our meals also plays a crucial role in weight management. A frequency-based dietary approach emphasizes the importance of consuming a balanced combination of macronutrients, including carbohydrates, proteins, and fats.

Carbohydrates provide the body with energy, and incorporating complex carbohydrates, such as whole grains and vegetables, can help promote satiety and prevent overeating. Proteins are essential for muscle repair and growth, and they also contribute to feelings of fullness. Including lean sources of protein, such as poultry, fish, and legumes, can support weight loss efforts.

Lastly, healthy fats, such as those found in avocados, nuts, and olive oil, are important for hormone regulation and overall well-being.

8.2.3 Mindful Eating and Portion Control

Another key aspect of frequency-based dietary approaches is practicing mindful eating and portion control. By paying attention to our body's hunger and fullness cues, we can avoid overeating and make more conscious food choices. This involves eating slowly, savoring each bite, and being aware of the body's signals of satisfaction.

Portion control is also essential for weight management. By consuming smaller, more frequent meals, individuals can naturally regulate their portion sizes and prevent excessive calorie intake. This approach allows for a more balanced distribution of energy throughout the day, which can help prevent energy crashes and cravings.

8.2.4 Intermittent Fasting

Intermittent fasting is a popular dietary approach that involves alternating periods of fasting and eating. This approach can be incorporated into a frequency-based weight management plan. By restricting the eating window to a specific time frame, such as 8 hours a day, individuals can naturally reduce their calorie intake and promote weight loss.

Intermittent fasting has been shown to have various benefits, including improved insulin sensitivity, increased fat burning, and reduced inflammation. However, it is important to note that this approach may not be suitable for everyone, and it is recommended to consult with a healthcare professional before implementing any fasting regimen.

8.2.5 The Role of Hydration

Hydration is often overlooked when it comes to weight management, but it plays a crucial role in overall health and well-being. Drinking an adequate amount of water throughout the day can help regulate appetite, support digestion, and enhance metabolism. Incorporating frequency-based reminders to drink water can ensure that individuals stay hydrated and support their weight management efforts.

8.2.6 Individualized Approaches

It is important to recognize that each individual is unique, and what works for one person may not work for another. Frequency-based dietary approaches should be tailored to an individual's specific needs, preferences, and lifestyle. Consulting with a registered dietitian or nutritionist can provide personalized guidance and support in developing a frequency-based weight management plan.

8.2.7 Long-Term Sustainability

Sustainable weight management is not just about short-term results; it is about adopting healthy habits that can be maintained over time. Frequency-based dietary approaches promote a balanced and mindful approach to eating, which can be sustained in the long run. By incorporating these strategies into daily life, individuals can not only achieve their weight management goals but also improve their overall well-being.

In conclusion, frequency-based dietary approaches for weight management focus on meal frequency, macronutrient balance, mindful eating, portion control, intermittent fasting, hydration, and individualization. By incorporating these strategies into our daily lives, we can optimize our weight management efforts and promote long-term well-being. Remember, it is essential to consult with a healthcare professional before making any significant changes to your diet or lifestyle.

8.3 Using Frequency to Boost Metabolism and Fat Burning

Frequency plays a crucial role in our body's metabolism and fat-burning processes. By understanding how frequency affects these mechanisms, we can harness its power to optimize weight loss efforts and achieve our desired goals. In this section, we will explore the ways in which frequency can be used to boost metabolism and enhance fat burning.

8.3.1 The Role of Frequency in Metabolism

Metabolism refers to the complex set of chemical reactions that occur within our bodies to convert food into energy. It is influenced by various factors, including genetics, age, and lifestyle choices. However, recent research has shown that frequency also plays a significant role in regulating metabolism.

8.3.2 Frequency's Impact on Metabolic Rate

Metabolic rate refers to the speed at which our bodies burn calories to produce energy. A higher metabolic rate means that we burn more calories even at rest, which can contribute to weight loss. Studies have found that certain frequencies can stimulate the body's metabolic rate, leading to increased calorie expenditure.

8.3.3 Using Frequency to Enhance Fat Burning

Fat burning is a key component of weight loss. When our bodies burn fat, they utilize stored energy reserves, leading to a reduction in body fat percentage. Frequency can be used to enhance fat burning by activating specific metabolic pathways that promote the breakdown of fat cells.

8.3.4 Frequency-Based Techniques for Boosting Metabolism

There are several frequency-based techniques that can be incorporated into our daily routines to boost metabolism and facilitate fat burning. These techniques include:

8.3.4.1 High-Intensity Interval Training (HIIT)

HIIT involves alternating between short bursts of intense exercise and periods of rest or low-intensity activity. Research has shown that incorporating specific frequencies during HIIT workouts can enhance fat burning and increase metabolic rate.

8.3.4.2 Cold Exposure

Exposing the body to cold temperatures, such as through cold showers or ice baths, can activate brown fat, a type of fat that burns calories to generate heat. By combining cold exposure with specific frequencies, we can further enhance the fat-burning effects.

8.3.4.3 Frequency-Based Dietary Approaches

Certain frequencies can also be incorporated into our dietary choices to boost metabolism and fat burning. For example, consuming foods that are rich in thermogenic compounds, such as chili peppers or green tea, can increase metabolic rate. By combining these foods with specific frequencies, we can maximize their effects.

8.3.5 Frequency Techniques for Overcoming Weight Loss Plateaus

Weight loss plateaus are common during a weight loss journey, where progress stalls despite continued efforts. Frequency can be a valuable tool for overcoming these plateaus and kickstarting the body's fat-burning processes once again.

8.3.5.1 Frequency Cycling

Frequency cycling involves periodically changing the frequencies used in our weight loss routines. This technique prevents the body from adapting to a specific frequency and helps to maintain a high metabolic rate.

8.3.5.2 Progressive Overload

Progressive overload is a principle commonly used in strength training, but it can also be applied to frequency-based weight loss strategies. By gradually increasing the intensity or duration of frequency-based exercises, we can continuously challenge our bodies and stimulate fat burning.

8.3.5.3 Combining Frequency with Other Weight Loss Strategies

Frequency can be combined with other proven weight loss strategies, such as calorie restriction and regular exercise, to enhance their effectiveness. By incorporating frequency techniques into a comprehensive weight loss plan, we can optimize our metabolism and maximize fat burning.

In conclusion, frequency has a significant impact on metabolism and fat burning. By understanding and utilizing the power of frequency, we can boost our metabolic rate, enhance fat burning, and overcome weight loss plateaus. Incorporating frequency-based techniques into our daily routines and combining them with other weight loss strategies can lead to more effective and sustainable weight loss results.

8.4 Frequency Techniques for Overcoming Weight Loss Plateaus

Weight loss plateaus can be frustrating and demotivating, especially when you have been diligently following a weight loss plan. However, by incorporating frequency techniques into your routine, you can overcome these plateaus and continue making progress towards your weight loss goals. In this section, we will explore various frequency techniques that can help you break through weight loss plateaus and achieve sustainable results.

8.4.1 Adjusting Frequency Patterns

One effective technique for overcoming weight loss plateaus is to adjust your frequency patterns. Our bodies have a natural tendency to adapt to repetitive stimuli, including dietary and exercise routines. By introducing variations in your frequency patterns, you can challenge your body and stimulate further weight loss.

One way to adjust frequency patterns is through intermittent fasting. This technique involves alternating periods of fasting with periods of eating. By changing the frequency of your meals, you can enhance your body's fat-burning capabilities and overcome weight loss plateaus. Additionally, intermittent fasting has been shown to improve insulin sensitivity and promote overall metabolic health.

8.4.2 Incorporating High-Intensity Interval Training (HIIT)

High-Intensity Interval Training (HIIT) is a popular exercise technique that involves short bursts of intense exercise followed by periods of rest or low-intensity exercise. By incorporating HIIT into your fitness routine, you can increase the frequency and intensity of your workouts, leading to enhanced weight loss.

HIIT has been shown to boost metabolism, increase fat burning, and improve cardiovascular health. By challenging your body with high-intensity intervals, you can break through weight loss plateaus and continue making progress towards your goals. Additionally, HIIT workouts are time-efficient, making them a convenient option for individuals with busy schedules.

8.4.3 Mindful Eating and Frequency

Mindful eating is a practice that involves paying attention to the present moment while consuming food. By incorporating frequency techniques into your mindful eating practice, you can develop a healthier relationship with food and overcome weight loss plateaus.

One technique is to eat slowly and savor each bite. By slowing down the frequency at which you eat, you can give your body ample time to register feelings of fullness, preventing overeating. Additionally, practicing mindful eating can help you become more attuned to your body's hunger and satiety cues, allowing you to make healthier food choices.

8.4.4 Incorporating Strength Training

Strength training is an essential component of any weight loss program. By incorporating frequency techniques into your strength training routine, you can overcome weight loss plateaus and build lean muscle mass.

One technique is to increase the frequency of your strength training sessions. Instead of focusing on long, infrequent workouts, consider breaking them into shorter, more frequent sessions. This approach can help you maintain an elevated metabolic rate throughout the day, leading to increased calorie burn and weight loss.

Additionally, incorporating compound exercises that target multiple muscle groups can further enhance the frequency and intensity of your workouts. Exercises such as squats, deadlifts, and bench presses engage multiple muscle groups simultaneously, leading to greater calorie expenditure and improved weight loss results.

8.4.5 Tracking and Adjusting Caloric Intake

Tracking and adjusting your caloric intake is crucial for overcoming weight loss plateaus. By monitoring your food intake and adjusting the frequency and quantity of your meals, you can create a calorie deficit necessary for weight loss.

One technique is to practice calorie cycling, which involves alternating between high and low-calorie days. On high-calorie days, you can increase your caloric intake to provide your body with the energy it needs for intense workouts. On low-calorie days, you can create a calorie deficit to promote weight loss. This cycling of calories can help prevent your body from adapting to a consistent caloric intake, leading to continued weight loss.

Additionally, adjusting macronutrient ratios can also be beneficial. Increasing protein intake can help preserve lean muscle mass and boost metabolism, while reducing carbohydrate intake can promote fat burning. Experimenting with different macronutrient ratios and adjusting them based on your body's response can help you overcome weight loss plateaus and achieve sustainable results.

8.4.6 Seeking Professional Guidance

If you find yourself struggling to overcome weight loss plateaus despite incorporating frequency techniques, it may be beneficial to seek professional guidance. A registered dietitian or a certified personal trainer can provide personalized recommendations based on your specific needs and goals. They can help you develop a tailored frequency-based weight loss plan and provide ongoing support and accountability.

Remember, weight loss plateaus are a normal part of the weight loss journey. By incorporating frequency techniques and staying consistent with your efforts, you can overcome these plateaus and continue making progress towards your desired weight and overall well-being.

In the next chapter, we will explore frequency techniques for managing stress and promoting overall mental health.

9 Frequency and Stress Management Techniques

9.1 Frequency-Based Stress Reduction Practices

Stress has become an increasingly prevalent issue in today's fast-paced and demanding world. It can have detrimental effects on both our physical and mental well-being if left unmanaged. Fortunately, frequency-based stress reduction practices offer a natural and effective way to alleviate anxiety and tension, promote relaxation, and enhance overall well-being.

The Power of Frequency in Stress Reduction

Frequency, in the context of stress reduction, refers to the specific vibrations or oscillations that can positively influence our physiological and psychological states. These frequencies can be harnessed through various techniques such as sound therapy, meditation, and mindfulness practices. By aligning our body and mind with specific frequencies, we can create a harmonious state that counteracts the negative effects of stress.

Sound Therapy for Stress Reduction

One of the most popular frequency-based stress reduction practices is sound therapy. Sound has been used for centuries as a healing modality, and its therapeutic effects on stress are well-documented. Different frequencies and tones can elicit specific responses in our brain and body, promoting relaxation, reducing anxiety, and improving overall well-being.

Sound therapy often involves the use of instruments such as singing bowls, tuning forks, or even recorded frequencies played through speakers or headphones. These instruments emit specific frequencies that resonate with different parts of our body, helping to release tension and restore balance. By immersing ourselves in these soothing sounds, we can create a calming environment that promotes stress reduction and relaxation.

Meditation and Mindfulness Practices

Meditation and mindfulness practices are another powerful way to harness the benefits of frequency for stress reduction. These practices involve focusing our attention on the present moment, cultivating a state of calm awareness, and letting go of stress-inducing thoughts and emotions.

During meditation, we can incorporate frequency-based techniques to enhance the relaxation response. This can be done by using guided meditations that incorporate specific frequencies or by silently repeating a calming mantra that resonates with a desired frequency. By immersing ourselves in these frequencies, we can create a deep sense of relaxation and inner peace.

Mindfulness practices, on the other hand, involve bringing our attention to the present moment without judgment. By tuning into the sensations in our body, the sounds around us, and the thoughts and emotions that arise, we can cultivate a state of mindfulness that helps us detach from stress and find a sense of calm. Incorporating frequency-based techniques, such as focusing on the breath and synchronizing it with specific frequencies, can further enhance the stress-reducing benefits of mindfulness.

Breathing Techniques and Frequency

Our breath is intimately connected to our stress response. When we are stressed, our breathing tends to become shallow and rapid, further exacerbating our feelings of anxiety and tension. By incorporating specific breathing techniques that align with frequency, we can activate the body's relaxation response and reduce stress.

One such technique is known as "resonant breathing" or "coherent breathing." This involves breathing in and out through the nose at a specific rhythm, typically around five to six breaths per minute. By synchronizing our breath with a specific frequency, we can create a coherent state in our body, promoting relaxation and reducing stress.

Frequency-Based Visualization and Imagery

Visualization and imagery techniques can also be enhanced by incorporating frequency. By creating vivid mental images that align with specific frequencies, we can tap into the power of our imagination to reduce stress and promote relaxation.

For example, during a guided visualization, we can imagine ourselves in a peaceful natural setting, such as a serene beach or a tranquil forest. As we immerse ourselves in this mental imagery, we can also visualize the corresponding frequency resonating throughout our body, bringing a sense of calm and tranquility. This combination of visualization and frequency can be a potent tool for stress reduction.

Incorporating Frequency into Daily Life

To fully reap the benefits of frequency-based stress reduction practices, it is essential to incorporate them into our daily lives. This can be done by setting aside dedicated time for sound therapy, meditation, or mindfulness practices. Additionally, we can integrate frequency into our daily routines by using frequency-based apps or recordings during moments of relaxation, such as before bed or during breaks throughout the day.

It is important to note that frequency-based stress reduction practices are not a one-size-fits-all solution. Each individual may respond differently to specific frequencies, so it is essential to explore and experiment with different techniques to find what works best for you. Whether it's through sound therapy, meditation, breathing techniques, or visualization, incorporating frequency into your stress reduction practices can have a profound impact on your overall well-being.

9.2 Using Frequency to Alleviate Anxiety and Tension

Anxiety and tension are common experiences that can significantly impact our overall well-being. They can manifest in various ways, such as racing thoughts, restlessness, irritability, and physical symptoms like increased heart rate and muscle tension. Fortunately, frequency can be a valuable tool in alleviating anxiety and tension, promoting a sense of calm and relaxation.

9.2.1 Understanding Anxiety and Tension

Anxiety is a natural response to stress or perceived threats. It is a normal part of life and can even be beneficial in certain situations, as it prepares us to face challenges. However, when anxiety becomes excessive or chronic, it can interfere with our daily functioning and overall quality of life.

Tension, on the other hand, refers to the physical and emotional strain that often accompanies anxiety. It can manifest as muscle tightness, headaches, digestive issues, and difficulty concentrating. Both anxiety and tension can be debilitating, making it crucial to find effective strategies to manage and alleviate them.

9.2.2 The Role of Frequency in Anxiety and Tension Reduction

Frequency-based techniques have shown promise in reducing anxiety and tension by influencing the body's physiological and psychological responses. These techniques involve exposing individuals to specific frequencies or vibrations that have a calming effect on the nervous system.

One such technique is binaural beats, which involve listening to two slightly different frequencies in each ear. The brain perceives the difference between the two frequencies as a rhythmic beat, which can induce a state of relaxation and reduce anxiety. Binaural beats have been found to promote the production of alpha and theta brainwaves, which are associated with relaxation and deep meditative states.

Another frequency-based technique is the use of music therapy. Certain types of music, such as classical or ambient music, have been found to have a soothing effect on the mind and body. The rhythmic patterns and harmonies in music can help regulate heart rate, breathing, and blood pressure, promoting a sense of calm and reducing anxiety and tension.

9.2.3 Frequency Techniques for Anxiety Relief

There are several frequency techniques that can be used to alleviate anxiety and tension:

9.2.3.1 Deep Breathing Exercises

Deep breathing exercises are a simple yet effective way to reduce anxiety and tension. By focusing on slow, deep breaths, you can activate the body's relaxation response and calm the nervous system. To enhance the effectiveness of deep breathing, you can incorporate specific frequencies or calming music into your practice. This can help synchronize your breath with the rhythm of the frequencies, further promoting relaxation.

9.2.3.2 Meditation and Mindfulness

Meditation and mindfulness practices involve focusing your attention on the present moment, cultivating a non-judgmental awareness of your thoughts and emotions. These practices have been shown to reduce anxiety and promote a sense of calm and well-being. Incorporating frequency-based techniques, such as binaural beats or calming music, can enhance the effectiveness of meditation and mindfulness by inducing a deeper state of relaxation.

9.2.3.3 Guided Imagery

Guided imagery involves using visualizations and mental imagery to create a sense of calm and relaxation. By imagining peaceful and serene scenes, you can shift your focus away from anxious thoughts and promote a sense of tranquility. Adding frequency-based elements, such as specific frequencies or music, can enhance the effectiveness of guided imagery and deepen the relaxation response.

9.2.3.4 Progressive Muscle Relaxation

Progressive muscle relaxation is a technique that involves systematically tensing and relaxing different muscle groups in the body. By consciously releasing tension from each muscle group, you can promote a state of physical and mental relaxation. Incorporating frequency-based techniques, such as binaural beats or calming music, can enhance the effectiveness of progressive muscle relaxation and deepen the relaxation response.

9.2.4 Frequency-Based Relaxation Methods

In addition to specific techniques, there are various frequency-based relaxation methods that can be incorporated into your daily routine to alleviate anxiety and tension:

9.2.4.1 Listening to Calming Music

Listening to calming music can have a profound impact on reducing anxiety and tension. Choose music with a slow tempo, gentle melodies, and soothing harmonies. Experiment with different genres, such as classical, ambient, or nature sounds, to find what resonates with you and promotes a sense of calm.

9.2.4.2 Nature Sounds and White Noise

Nature sounds, such as ocean waves, rain, or birdsong, can create a peaceful and relaxing environment. Similarly, white noise, such as the sound of a fan or a gentle hum, can help mask distracting noises and promote a sense of tranquility. These frequency-based sounds can be used during meditation, sleep, or any time you need to create a calm atmosphere.

9.2.4.3 Frequency-Based Apps and Devices

There are numerous apps and devices available that utilize frequency-based techniques to promote relaxation and reduce anxiety. These tools often incorporate binaural beats, calming music, or guided meditations to induce a state of calm. Explore different options and find the ones that resonate with you and fit into your lifestyle.

9.2.4.4 Yoga and Tai Chi

Yoga and Tai Chi are ancient practices that combine gentle movements, breath control, and mindfulness. These practices have been shown to reduce anxiety and tension by promoting relaxation, improving body awareness, and calming the mind. Incorporating frequency-based elements, such as calming music or specific frequencies, can enhance the relaxation and stress-reducing benefits of these practices.

Incorporating frequency-based techniques and relaxation methods into your daily routine can help alleviate anxiety and tension, promoting a greater sense of calm and well-being. Experiment with different techniques and find what works best for you. Remember, consistency is key, so make frequency-based practices a regular part of your self-care routine.

9.3 Frequency Techniques for Stress Resilience and Coping

Stress is an inevitable part of life, and its impact on our well-being cannot be underestimated. The constant demands and pressures we face can take a toll on our mental and physical health. However, by understanding the power of frequency and utilizing specific techniques, we can build resilience and effectively cope with stress.

9.3.1 The Role of Frequency in Stress Resilience

Frequency plays a crucial role in stress resilience by influencing our brainwaves and physiological responses. Different frequencies have distinct effects on our mental and emotional states, allowing us to regulate stress levels and promote a sense of calm and balance.

9.3.2 Harnessing the Power of Frequency for Stress Reduction

There are various frequency techniques that can be employed to reduce stress and enhance resilience. These techniques leverage the natural rhythms of our brain and body to induce relaxation and restore equilibrium. Let's explore some of these techniques:

9.3.2.1 Binaural Beats

Binaural beats are a popular frequency technique used for stress reduction. They involve listening to two slightly different frequencies in each ear, which creates a third frequency in the brain. This third frequency corresponds to a specific brainwave state, such as alpha or theta, which promotes relaxation and reduces stress.

To experience the benefits of binaural beats, one can use headphones and listen to specially designed audio tracks or apps that provide different frequencies for specific purposes, such as stress reduction or deep relaxation.

9.3.2.2 Isochronic Tones

Similar to binaural beats, isochronic tones are another effective frequency technique for stress resilience. These tones involve the use of evenly spaced pulses of sound or light, which entrain the brain to a desired frequency. By listening to or watching these tones, individuals can induce a state of relaxation and reduce stress.

Isochronic tones can be found in various forms, including audio tracks, videos, or specialized devices. They can be incorporated into meditation practices or used as standalone tools for stress reduction.

9.3.2.3 Solfeggio Frequencies

Solfeggio frequencies are a set of ancient musical frequencies that have been associated with numerous health benefits, including stress reduction. Each frequency in the Solfeggio scale is believed to have specific healing properties and can be used to balance and harmonize the mind and body.

Listening to music or tones based on Solfeggio frequencies can help alleviate stress and promote a sense of well-being. These frequencies can be found in various forms, such as music tracks, sound baths, or even incorporated into guided meditation practices.

9.3.2.4 Resonance Frequency Breathing

Resonance frequency breathing is a technique that involves breathing at a specific rate to synchronize the heart and respiratory systems. By breathing at a specific frequency, typically around six breaths per minute, individuals can activate the body's relaxation response and reduce stress.

To practice resonance frequency breathing, one can sit comfortably and inhale for a count of four, hold the breath for a count of four, and exhale for a count of four. This rhythmic breathing pattern can be repeated for several minutes, allowing the body to enter a state of deep relaxation.

9.3.3 Integrating Frequency Techniques into Daily Life

To effectively manage stress and build resilience, it is essential to integrate frequency techniques into our daily lives. Here are some practical tips for incorporating these techniques:

9.3.3.1 Morning and Evening Rituals

Start and end your day with a frequency-based practice. Whether it's listening to binaural beats or practicing resonance frequency breathing, dedicating a few minutes in the morning and evening can set the tone for a calm and balanced day.

9.3.3.2 Mindful Breaks

Take short breaks throughout the day to engage in a frequency technique of your choice. This could involve listening to isochronic tones during a lunch break or practicing deep breathing exercises during a stressful moment. These mindful breaks can help reset your stress response and improve overall well-being.

9.3.3.3 Incorporating Frequency into Exercise

Combine frequency techniques with physical activity to enhance stress resilience. For example, listen to Solfeggio frequencies while practicing yoga or incorporate binaural beats into your running routine. This integration can amplify the benefits of both frequency and exercise, promoting a greater sense of well-being.

9.3.3.4 Creating a Frequency Sanctuary

Designate a specific space in your home as a frequency sanctuary. Fill this space with calming elements, such as candles, essential oils, and comfortable seating. Use this sanctuary for your frequency practices, allowing it to become a haven for stress reduction and relaxation.

By incorporating these frequency techniques into our daily lives, we can effectively manage stress, build resilience, and enhance our overall well-being. The power of frequency is a valuable tool that can help us navigate the challenges of life with greater ease and balance.

9.4 Frequency-Based Relaxation Methods

Relaxation is a vital component of overall well-being and plays a significant role in managing stress, improving sleep quality, and promoting a positive mindset. In this section, we will explore various frequency-based relaxation methods that can help you achieve a state of deep relaxation and enhance your overall well-being.

9.4.1 Binaural Beats

Binaural beats are a popular form of frequency-based relaxation technique that involves listening to two slightly different frequencies in each ear. The brain perceives the difference between these frequencies as a rhythmic beat, which can induce a state of relaxation and calmness. Binaural beats have been found to promote relaxation, reduce anxiety, and improve sleep quality. By listening to specific frequencies, you can target different states of relaxation, such as alpha waves for relaxation and creativity or delta waves for deep sleep.

9.4.2 Isochronic Tones

Similar to binaural beats, isochronic tones are another form of frequency-based relaxation method. Instead of using two different frequencies, isochronic tones use a single tone that turns on and off at a specific rhythm. This rhythmic pattern helps synchronize the brainwaves and induce a state of relaxation. Isochronic tones have been shown to reduce stress, improve focus, and enhance overall well-being. They can be used during meditation, before sleep, or anytime you need to relax and unwind.

9.4.3 Guided Imagery

Guided imagery is a relaxation technique that involves using visualizations and mental imagery to promote relaxation and reduce stress. By listening to a guided meditation or visualization, you can create a mental image of a peaceful and calming place, such as a beach or a forest. This technique helps shift your focus away from stress and negative thoughts, allowing you to enter a state of deep relaxation. Guided imagery can be combined with frequency-based techniques, such as binaural beats or isochronic tones, to enhance the relaxation experience.

9.4.4 Progressive Muscle Relaxation

Progressive muscle relaxation is a technique that involves systematically tensing and relaxing different muscle groups in the body. By consciously tensing and then releasing the tension in each muscle group, you can promote a deep sense of relaxation and release physical tension. This technique can be combined with frequency-based methods, such as listening to relaxing music or nature sounds, to enhance the relaxation response. Progressive muscle relaxation is particularly effective for reducing muscle tension, relieving stress, and improving sleep quality.

9.4.5 Deep Breathing Exercises

Deep breathing exercises are a simple yet powerful relaxation technique that can be practiced anywhere, at any time. By focusing on your breath and taking slow, deep breaths, you can activate the body's relaxation response and reduce stress. Deep breathing helps slow down the heart rate, lower blood pressure, and promote a sense of calmness. You can enhance the effects of deep breathing exercises by incorporating frequency-based techniques, such as listening to calming music or nature sounds.

9.4.6 Yoga and Tai Chi

Yoga and Tai Chi are ancient practices that combine physical movement, breath control, and mindfulness to promote relaxation and overall well-being. These practices incorporate gentle movements, stretching, and meditation, which help release tension, improve flexibility, and calm the mind. By incorporating frequency-based techniques, such as listening to soothing music or using binaural beats, you can deepen the relaxation experience during yoga or Tai Chi sessions.

9.4.7 Aromatherapy

Aromatherapy is a relaxation technique that involves using essential oils to promote relaxation and reduce stress. Certain essential oils, such as lavender, chamomile, and bergamot, have been found to have calming and soothing effects on the mind and body. By diffusing these oils or using them in massage oils or bath products, you can create a relaxing environment and enhance the relaxation response. Combining aromatherapy with frequency-based techniques, such as listening to relaxing music or using isochronic tones, can amplify the relaxation benefits.

9.4.8 Mindfulness Meditation

Mindfulness meditation is a practice that involves focusing your attention on the present moment without judgment. By cultivating a state of mindfulness, you can reduce stress, enhance self-awareness, and promote relaxation. Mindfulness meditation can be combined with frequency-based techniques, such as listening to guided meditations or using binaural beats, to deepen the relaxation experience and promote a sense of calmness.

Incorporating frequency-based relaxation methods into your daily routine can have profound effects on your overall well-being. Whether you choose to listen to binaural beats, practice deep breathing exercises, or engage in yoga and Tai Chi, these techniques can help you manage stress, improve sleep quality, and cultivate a positive mindset. Experiment with different methods and find what works best for you. Remember, relaxation is a journey, and by incorporating frequency-based relaxation methods into your life, you can experience the transformative power of relaxation on your well-being.

10 Frequency and Mental Health

10.1 Frequency's Impact on Mental Well-Being

Frequency, the measure of how often something occurs, has a profound impact on our mental well-being. The vibrations and patterns of frequency can influence our brain waves, emotions, and cognitive function. In this section, we will explore the various ways in which frequency affects our mental health and well-being.

10.1.1 The Power of Frequency in Shaping Mental Well-Being

The human brain operates on different frequencies, known as brain waves, which are associated with different states of consciousness. These brain waves include delta, theta, alpha, beta, and gamma waves. Each frequency range corresponds to specific mental states and functions.

Research has shown that certain frequencies can have a direct impact on our mental well-being. For example, alpha waves, which are associated with relaxation and a calm state of mind, can be induced through frequency-based techniques such as meditation or listening to soothing music. These techniques can help reduce anxiety, promote relaxation, and improve overall mental well-being.

10.1.2 Using Frequency to Manage Depression and Mood Disorders

Depression and mood disorders are prevalent mental health conditions that can significantly impact an individual's quality of life. Frequency-based therapies have shown promise in managing these conditions and improving mental well-being.

Studies have found that specific frequencies can stimulate the release of neurotransmitters such as serotonin and dopamine, which play a crucial role in regulating mood. By targeting these frequencies, individuals experiencing depression or mood disorders may find relief and experience an improvement in their overall mental well-being.

10.1.3 Frequency-Based Strategies for Anxiety Relief

Anxiety is another common mental health condition that can have a significant impact on an individual's well-being. Frequency-based strategies can be effective in reducing anxiety and promoting a sense of calm.

One such strategy is the use of binaural beats, which involve listening to two slightly different frequencies in each ear. This creates a third frequency in the brain, known as the binaural beat, which can induce a state of relaxation and reduce anxiety. Research has shown that binaural beats can help alleviate symptoms of anxiety and improve overall mental well-being.

10.1.4 Frequency Techniques for Enhancing Cognitive Function

Cognitive function, including memory, attention, and problem-solving abilities, is essential for overall mental well-being. Frequency-based techniques can be used to enhance cognitive function and improve mental clarity.

For example, gamma waves, which are associated with increased focus and cognitive processing, can be stimulated through frequency-based practices such as brainwave entrainment. This technique involves exposing the brain to specific frequencies, which can enhance cognitive function and improve mental well-being.

10.1.5 The Role of Frequency in Mindfulness and Present Moment Awareness

Mindfulness, the practice of being fully present in the moment, has been shown to have numerous benefits for mental well-being. Frequency can play a role in cultivating mindfulness and promoting present moment awareness.

By using frequency-based techniques such as guided meditation or sound therapy, individuals can enhance their ability to focus on the present moment and reduce distractions. This can lead to a greater sense of well-being, improved mental clarity, and reduced stress levels.

10.1.6 Frequency's Influence on Emotional Well-Being

Emotions play a significant role in our mental well-being, and frequency can have a profound impact on our emotional state. Different frequencies can evoke specific emotions and influence our overall emotional well-being.

For example, listening to music with a slow tempo and low-frequency sounds can induce a sense of calm and relaxation. On the other hand, music with a fast tempo and high-frequency sounds can evoke feelings of excitement and energy. By understanding the influence of frequency on emotions, individuals can use this knowledge to regulate their emotional well-being and promote a positive mindset.

10.1.7 Frequency-Based Techniques for Overall Mental Well-Being

Incorporating frequency-based techniques into our daily lives can have a significant impact on our mental well-being. Here are some strategies that can be used to enhance overall mental well-being:

1. Meditation: Practicing meditation with frequency-based techniques such as binaural beats or guided meditation can promote relaxation, reduce stress, and improve mental clarity.
2. Sound therapy: Listening to specific frequencies or music designed to induce a desired mental state can help regulate emotions, reduce anxiety, and enhance overall mental well-being.
3. Brainwave entrainment: Using audio recordings or devices that emit specific frequencies can entrain the brain to desired states, such as increased focus or relaxation.
4. Mindfulness practices: Incorporating frequency-based techniques into mindfulness practices can enhance present moment awareness, reduce distractions, and promote overall mental well-being.

By incorporating these frequency-based techniques into our daily routines, we can harness the power of frequency to improve our mental well-being and enhance our overall quality of life.

In the next section, we will explore the relationship between frequency and physical health, including its influence on physical performance, pain management, and immune system function.

10.2 Using Frequency to Manage Depression and Mood Disorders

Depression and mood disorders are prevalent mental health conditions that can significantly impact an individual's overall well-being. These conditions can manifest as persistent feelings of sadness, hopelessness, and a loss of interest in activities once enjoyed. Fortunately, research has shown that frequency can play a crucial role in managing and alleviating symptoms of depression and mood disorders.

10.2.1 Understanding Depression and Mood Disorders

Depression is a complex mental health condition characterized by persistent feelings of sadness, a lack of motivation, and a general disinterest in life. It can affect individuals of all ages and can have a profound impact on their daily functioning and quality of life. Mood disorders, on the other hand, encompass a broader range of conditions, including bipolar disorder, major depressive disorder, and seasonal affective disorder (SAD).

10.2.2 The Role of Frequency in Managing Depression

Research has shown that frequency can have a significant impact on managing depression and mood disorders. The brain operates on different frequencies, and imbalances in these frequencies can contribute to the development and exacerbation of depressive symptoms. By using specific frequencies, individuals can stimulate the brain to promote a more balanced and positive mental state.

10.2.3 Frequency-Based Strategies for Managing Depression

There are several frequency-based strategies that can be employed to manage depression and mood disorders effectively. One such strategy is the use of binaural beats, which involve listening to two slightly different frequencies in each ear. This technique has been found to promote relaxation, reduce anxiety, and improve mood.

Another effective strategy is the use of isochronic tones, which are repetitive beats of a single frequency. These tones can help induce a state of deep relaxation and calmness, reducing the symptoms of depression and promoting a more positive mindset.

10.2.4 Frequency Techniques for Alleviating Anxiety

Anxiety often coexists with depression and mood disorders, and managing anxiety symptoms is crucial for overall well-being. Frequency-based techniques can be highly effective in alleviating anxiety and promoting a sense of calmness. For example, alpha waves, which have a frequency range of 8 to 12 Hz, can help reduce anxiety and induce a state of relaxation.

Similarly, theta waves, with a frequency range of 4 to 8 Hz, can promote deep relaxation and reduce anxiety levels. By incorporating these frequencies into daily practices such as meditation or listening to specially designed audio tracks, individuals can experience a significant reduction in anxiety symptoms.

10.2.5 Frequency Techniques for Enhancing Cognitive Function

Depression and mood disorders can often lead to cognitive impairments, including difficulties with concentration, memory, and decision-making.

Frequency-based techniques can help enhance cognitive function and improve mental clarity.

Gamma waves, with a frequency range of 25 to 100 Hz, have been associated with increased cognitive abilities, including improved memory and focus. By incorporating gamma wave stimulation through techniques such as brainwave entrainment or neurofeedback, individuals can experience enhanced cognitive function and improved overall mental well-being.

10.2.6 Seeking Professional Guidance

While frequency-based techniques can be beneficial in managing depression and mood disorders, it is essential to seek professional guidance and support. Mental health professionals, such as therapists or psychiatrists, can provide personalized treatment plans and guidance tailored to individual needs.

These professionals can help individuals navigate the complexities of their mental health conditions and provide additional strategies and interventions to complement frequency-based techniques. It is crucial to approach the management of depression and mood disorders holistically, combining frequency-based practices with evidence-based therapies and interventions.

10.2.7 Conclusion

Frequency plays a significant role in managing depression and mood disorders. By utilizing specific frequencies, individuals can stimulate the brain to promote a more balanced and positive mental state. Frequency-based strategies, such as binaural beats and isochronic tones, can be highly effective in managing symptoms of depression and anxiety, promoting relaxation, and enhancing cognitive function.

However, it is important to remember that frequency-based techniques should be used in conjunction with professional guidance and support. Mental health professionals can provide personalized treatment plans and additional interventions to complement frequency-based practices. By taking a holistic approach to mental health, individuals can effectively manage their depression and mood disorders, leading to improved overall well-being.

10.3 Frequency-Based Strategies for Anxiety Relief

Anxiety is a common mental health condition that affects millions of people worldwide. It can manifest as persistent worry, fear, and unease, often accompanied by physical symptoms such as rapid heartbeat, sweating, and restlessness. While there are various treatment options available for anxiety, including therapy and medication, emerging research suggests that frequency-based strategies can also play a significant role in anxiety relief.

10.3.1 Understanding Anxiety and its Relationship with Frequency

Anxiety is a complex condition that can be influenced by multiple factors, including genetics, environment, and lifestyle. One such factor that has gained attention in recent years is the role of frequency in anxiety management. Frequency refers to the rate at which something occurs or vibrates, and it has been found to have a profound impact on our mental and emotional well-being.

10.3.2 The Science Behind Frequency and Anxiety Relief

Research has shown that different frequencies can have specific effects on the brain and nervous system, which in turn can influence our emotional state. For example, low-frequency sounds, such as those found in nature or certain types of music, have been found to promote relaxation and reduce anxiety. On the other hand, high-frequency sounds, such as alarms or loud noises, can trigger a stress response and increase feelings of anxiety.

10.3.3 Using Frequency to Regulate the Nervous System

One of the ways in which frequency-based strategies can help alleviate anxiety is by regulating the autonomic nervous system (ANS). The ANS is responsible for controlling our body's involuntary functions, including heart rate, blood pressure, and digestion. When we experience anxiety, the sympathetic branch of the ANS, also known as the "fight-or-flight" response, becomes activated, leading to increased arousal and heightened anxiety symptoms.

By utilizing specific frequencies, such as those found in certain types of music or sound therapy, it is possible to stimulate the parasympathetic branch of the ANS, which promotes relaxation and counteracts the effects of the sympathetic response. This can help reduce anxiety symptoms and create a sense of calm and well-being.

10.3.4 Frequency-Based Techniques for Anxiety Relief

There are several frequency-based techniques that can be incorporated into daily life to help manage anxiety. Here are a few examples:

10.3.4.1 Binaural Beats

Binaural beats are a type of auditory illusion created by playing two slightly different frequencies in each ear. This creates a third frequency, known as the binaural beat, which the brain perceives as a rhythmic pulsation. Research suggests that listening to binaural beats in the alpha or theta frequency range can promote relaxation and reduce anxiety.

10.3.4.2 Guided Imagery

Guided imagery involves using visualizations and mental imagery to create a sense of calm and relaxation. By combining soothing imagery with specific frequencies, such as those found in nature sounds or calming music, guided imagery can help reduce anxiety and promote a sense of inner peace.

10.3.4.3 Sound Bath Therapy

Sound bath therapy involves immersing oneself in a bath of sound vibrations created by various instruments, such as singing bowls, gongs, and tuning forks. The frequencies produced by these instruments can help induce a state of deep relaxation and release tension, thereby reducing anxiety.

10.3.4.4 Mindfulness Meditation

Mindfulness meditation involves focusing one's attention on the present moment without judgment. By incorporating specific frequencies, such as those found in meditation music or chanting, mindfulness meditation can help calm the mind, reduce anxiety, and promote overall well-being.

10.3.5 Integrating Frequency-Based Strategies into Daily Life

Incorporating frequency-based strategies for anxiety relief into daily life can be a powerful tool for managing anxiety. Here are some practical tips for integrating these strategies:

- Create a designated relaxation space in your home where you can engage in frequency-based practices without distractions.
- Set aside dedicated time each day for frequency-based activities, such as listening to calming music or engaging in guided imagery.
- Experiment with different frequencies and techniques to find what works best for you. Everyone's response to frequency can vary, so it's important to explore and discover what resonates with you.
- Consider incorporating frequency-based practices into your bedtime routine to promote relaxation and improve sleep quality.

- Seek guidance from a qualified professional, such as a sound therapist or meditation teacher, to learn more about specific frequency-based techniques and how to incorporate them effectively.

By incorporating frequency-based strategies for anxiety relief into your daily routine, you can empower yourself to take an active role in managing your anxiety and promoting overall well-being. Remember, finding the right frequency for you may take time and experimentation, but the potential benefits are well worth the effort.

10.4 Frequency Techniques for Enhancing Cognitive Function

Cognitive function refers to the mental processes and abilities that allow us to think, reason, learn, and remember. It encompasses various aspects such as attention, memory, problem-solving, decision-making, and creativity. The impact of frequency on cognitive function is an area of growing interest and research. In this section, we will explore the different frequency techniques that can enhance cognitive function and improve overall mental performance.

10.4.1 Brainwave Entrainment

One of the most widely studied frequency techniques for enhancing cognitive function is brainwave entrainment. Brainwave entrainment involves the use of external stimuli, such as sound or light, to synchronize the brainwaves to a specific frequency. This technique has been found to have a profound impact on cognitive abilities.

Different frequencies are associated with different states of consciousness and cognitive functions. For example, the alpha frequency range (8-12 Hz) is associated with relaxed and focused states, while the beta frequency range (12-30 Hz) is linked to alertness and concentration. By exposing the brain to specific frequencies, brainwave entrainment can help induce desired cognitive states.

Research has shown that brainwave entrainment can improve attention, memory, and overall cognitive performance. It has been used to enhance learning abilities, increase creativity, and improve problem-solving skills. By entraining the brain to specific frequencies, individuals can optimize their cognitive function and achieve peak mental performance.

10.4.2 Binaural Beats

Binaural beats are another popular frequency technique used to enhance cognitive function. Binaural beats are created by playing two slightly different frequencies in each ear, which the brain then perceives as a single beat. This beat corresponds to the difference between the two frequencies and can influence brainwave activity.

Similar to brainwave entrainment, binaural beats can help induce specific cognitive states. For example, playing binaural beats in the alpha frequency range can promote relaxation and improve focus. On the other hand, binaural beats in the beta frequency range can increase alertness and enhance concentration.

Studies have shown that listening to binaural beats can improve cognitive performance in various domains. It has been found to enhance memory, attention, and problem-solving abilities. Binaural beats have also been used to reduce anxiety and improve mood, further enhancing cognitive function.

10.4.3 Meditation and Mindfulness

Meditation and mindfulness practices have long been associated with cognitive benefits. These practices involve focusing attention and cultivating present moment awareness. Frequency techniques can be incorporated into meditation and mindfulness practices to enhance their cognitive effects.

By using specific frequencies during meditation, individuals can deepen their state of relaxation and focus. For example, playing soothing frequencies in the theta range (4-8 Hz) can promote a deep meditative state and enhance introspection. This can lead to improved cognitive function, including enhanced creativity and problem-solving abilities.

Mindfulness practices, which involve non-judgmental awareness of the present moment, can also be enhanced with frequency techniques. By incorporating binaural beats or other frequency-based stimuli, individuals can deepen their mindfulness practice and improve cognitive function.

10.4.4 Neurofeedback Training

Neurofeedback training is a technique that uses real-time feedback of brainwave activity to train individuals to self-regulate their brain function. It involves measuring brainwave activity using sensors placed on the scalp and providing feedback through visual or auditory cues.

By using frequency-based feedback, neurofeedback training can help individuals improve their cognitive function. For example, if an individual is experiencing difficulties with attention, neurofeedback training can provide feedback when their brainwaves are in the desired frequency range associated with focused attention. Over time, this can help train the brain to achieve and maintain the desired cognitive state.

Neurofeedback training has been used to enhance cognitive function in various populations, including individuals with attention deficit hyperactivity disorder (ADHD) and traumatic brain injury (TBI). It has been found to improve attention, memory, and executive function.

10.4.5 Cognitive Training Programs

In addition to frequency techniques, cognitive training programs can also be used to enhance cognitive function. These programs involve engaging in specific mental exercises and activities designed to target and improve different cognitive abilities.

Frequency techniques can be integrated into cognitive training programs to enhance their effectiveness. For example, incorporating brainwave entrainment or binaural beats into cognitive training exercises can help optimize brain function and improve cognitive performance.

Cognitive training programs have been found to improve various aspects of cognitive function, including attention, memory, and problem-solving abilities. They have been used to enhance cognitive performance in healthy individuals as well as those with cognitive impairments.

In conclusion, frequency techniques offer promising ways to enhance cognitive function and improve mental performance. Brainwave entrainment, binaural beats, meditation and mindfulness, neurofeedback training, and cognitive training programs are all effective methods for optimizing cognitive abilities. By incorporating these techniques into daily routines, individuals can unlock their full cognitive potential and experience improved mental clarity, focus, and overall cognitive well-being.

11 Frequency and Physical Health

11.1 Frequency's Influence on Physical Performance and Endurance

Frequency, as a fundamental aspect of our existence, has a profound impact on various aspects of our lives, including physical performance and endurance. In this section, we will explore how frequency influences our body's ability to perform and endure physical activities, and how we can harness its power to optimize our physical health.

The Role of Frequency in Physical Performance

Physical performance refers to the ability of our body to carry out tasks and activities efficiently and effectively. Whether it's participating in sports, engaging in exercise routines, or simply performing daily activities, our physical performance plays a crucial role in our overall well-being. Frequency, with its ability to influence our body at a cellular level, has a significant impact on our physical performance.

Cellular Resonance and Energy Transfer

At the core of frequency's influence on physical performance lies the concept of cellular resonance. Every cell in our body has its own unique frequency, and when these frequencies are in harmony, they create a state of coherence that enhances cellular communication and energy transfer. This coherence allows for optimal functioning of our body's systems, leading to improved physical performance.

Enhancing Muscle Function and Strength

Frequency has been found to have a direct impact on muscle function and strength. When exposed to specific frequencies, our muscles can experience increased blood flow, oxygenation, and nutrient delivery, resulting in improved muscle performance. Additionally, frequency-based techniques, such as vibration therapy, have been shown to enhance muscle activation and recruitment, leading to increased strength and power.

Improving Endurance and Stamina

Endurance, the ability to sustain physical activity over an extended period, is another area where frequency plays a crucial role. By optimizing cellular communication and energy transfer, frequency can enhance our body's ability to efficiently utilize energy sources, such as glucose and fatty acids, during prolonged physical exertion. This improved energy utilization translates into increased endurance and stamina, allowing us to perform at higher levels for longer durations.

Frequency-Based Approaches for Physical Performance Enhancement

Now that we understand the influence of frequency on physical performance, let's explore some practical approaches to harness its power for optimal physical health.

Frequency-Specific Training

Frequency-specific training involves the use of specific frequencies to target and stimulate different physiological systems in the body. By incorporating frequency-specific exercises and techniques into our training routines, we can enhance muscle activation, improve coordination, and optimize energy utilization. This targeted approach allows us to maximize the benefits of frequency on physical performance.

Vibration Therapy

Vibration therapy, a popular frequency-based technique, involves the use of vibrating platforms or devices to stimulate muscle contractions and improve circulation. This therapy has been shown to enhance muscle strength, flexibility, and balance, making it an effective tool for physical performance enhancement. Incorporating vibration therapy into our exercise routines can help optimize muscle function and improve overall physical performance.

Frequency-Based Recovery Strategies

Recovery is an essential component of physical performance, as it allows our body to repair and rebuild after intense physical activity. Frequency-based recovery strategies, such as using specific frequencies for relaxation and tissue regeneration, can aid in reducing muscle soreness, inflammation, and fatigue. By incorporating these strategies into our post-workout routines, we can enhance our body's ability to recover and improve overall physical performance.

The Impact of Frequency on Endurance Sports

Endurance sports, such as long-distance running, cycling, and swimming, require sustained physical effort over extended periods. Frequency's influence on endurance sports is particularly significant, as it can directly impact our body's ability to endure and perform at high levels.

Optimizing Energy Utilization

Endurance sports heavily rely on our body's ability to efficiently utilize energy sources. By optimizing cellular communication and energy transfer through frequency-based techniques, we can enhance our body's ability to utilize glucose and fatty acids as fuel sources. This improved energy utilization can delay the onset of fatigue and improve overall endurance performance.

Enhancing Oxygenation and Blood Flow

Oxygenation and blood flow are crucial factors in endurance sports, as they directly impact our body's ability to deliver oxygen and nutrients to working muscles. Frequency-based techniques, such as specific breathing exercises and targeted frequency exposure, can enhance oxygenation and blood flow, improving endurance performance and delaying the onset of fatigue.

Mental Resilience and Focus

Endurance sports not only require physical stamina but also mental resilience and focus. Frequency-based techniques, such as binaural beats and brainwave entrainment, can help induce states of relaxation, focus, and mental clarity. By incorporating these techniques into our training and competition routines, we can enhance our mental resilience, maintain focus, and improve overall endurance performance.

In conclusion, frequency's influence on physical performance and endurance is undeniable. By understanding and harnessing the power of frequency, we can optimize our body's ability to perform physical tasks efficiently, improve endurance, and enhance overall physical health. Incorporating frequency-based approaches into our training, recovery, and endurance sports routines can unlock our full potential and lead to a more fulfilling and rewarding physical experience.

11.2 Using Frequency for Pain Management and Rehabilitation

Pain is a common experience that can significantly impact our daily lives and overall well-being. Whether it is acute or chronic, pain can limit our ability to perform daily activities, affect our mood, and decrease our quality of life. Traditional approaches to pain management often involve medication or physical therapy, but there is growing evidence that frequency-based techniques can also be effective in alleviating pain and promoting rehabilitation.

11.2.1 Understanding the Role of Frequency in Pain Management

Frequency, in the context of pain management, refers to the specific vibrations or oscillations that are applied to the body. These vibrations can be delivered through various methods such as sound waves, electrical stimulation, or mechanical devices. The underlying principle is that these frequencies can interact with the body's cells, tissues, and nervous system, influencing pain perception and promoting healing.

11.2.2 The Science Behind Frequency and Pain Relief

Research has shown that frequency-based therapies can have a profound impact on pain perception and management. One of the key mechanisms through which frequency influences pain is by stimulating the release of endorphins, which are natural pain-relieving chemicals produced by the body. By increasing endorphin levels, frequency-based therapies can help reduce pain intensity and improve overall comfort.

Additionally, frequency-based therapies have been found to modulate the activity of the nervous system. By targeting specific frequencies, these therapies can help regulate the transmission of pain signals, effectively reducing the perception of pain. This modulation of the nervous system can also promote relaxation and decrease muscle tension, further contributing to pain relief.

11.2.3 Frequency-Based Techniques for Pain Management

There are several frequency-based techniques that can be used for pain management and rehabilitation. These techniques can be applied in various settings, including clinical settings, rehabilitation centers, or even in the comfort of your own home. Some commonly used frequency-based techniques for pain management include:

11.2.3.1 Transcutaneous Electrical Nerve Stimulation (TENS)

TENS is a non-invasive technique that involves the use of low-frequency electrical currents to stimulate the nerves and provide pain relief. The electrical currents are delivered through electrodes placed on the skin near the site of pain. TENS has been found to be effective in managing various types of pain, including musculoskeletal pain, neuropathic pain, and postoperative pain.

11.2.3.2 Low-Level Laser Therapy (LLLT)

LLLT, also known as cold laser therapy, utilizes low-intensity laser light to stimulate cellular activity and promote tissue healing. The laser light is applied directly to the affected area, penetrating the skin and targeting the underlying tissues. LLLT has been shown to be effective in reducing pain and inflammation, accelerating tissue repair, and improving overall function.

11.2.3.3 Vibroacoustic Therapy

Vibroacoustic therapy involves the use of low-frequency vibrations and sound waves to stimulate the body's tissues and promote relaxation. This therapy can be delivered through specialized devices that transmit vibrations to specific areas of the body or through specially designed furniture, such as vibroacoustic beds or chairs. Vibroacoustic therapy has been found to be beneficial in managing chronic pain, reducing muscle tension, and improving sleep quality.

11.2.4 Frequency-Based Rehabilitation Techniques

In addition to pain management, frequency-based techniques can also play a crucial role in the rehabilitation process. Whether recovering from an injury, surgery, or a chronic condition, rehabilitation aims to restore function, improve mobility, and enhance overall well-being. Frequency-based rehabilitation techniques can complement traditional rehabilitation approaches and accelerate the healing process.

11.2.4.1 Ultrasound Therapy

Ultrasound therapy utilizes high-frequency sound waves to generate heat within the body's tissues. This heat can increase blood flow, promote tissue healing, and reduce pain and inflammation. Ultrasound therapy is commonly used in the rehabilitation of musculoskeletal injuries, such as sprains, strains, and tendonitis.

11.2.4.2 Electrical Muscle Stimulation (EMS)

EMS involves the use of electrical currents to stimulate muscle contractions. This technique can be used to strengthen weakened muscles, improve range of motion, and enhance overall muscle function. EMS is often employed in the rehabilitation of orthopedic injuries, neurological conditions, and post-surgical recovery.

11.2.4.3 Whole Body Vibration (WBV)

WBV therapy involves standing, sitting, or lying on a vibrating platform that transmits low-frequency vibrations throughout the body. These vibrations stimulate muscle contractions and promote neuromuscular activation. WBV has been shown to improve muscle strength, balance, and flexibility, making it a valuable tool in the rehabilitation process.

11.2.5 The Benefits of Frequency-Based Pain Management and Rehabilitation

The use of frequency-based techniques for pain management and rehabilitation offers several benefits. These techniques are non-invasive, drug-free, and generally well-tolerated, making them suitable for a wide range of individuals. Additionally, frequency-based therapies can be used in conjunction with other treatment modalities, enhancing their effectiveness and promoting holistic healing.

Furthermore, frequency-based techniques have been found to have minimal side effects and can be easily incorporated into daily routines. Whether used as a standalone therapy or as part of a comprehensive pain management or rehabilitation program, frequency-based techniques provide individuals with additional tools to manage their pain, improve their function, and enhance their overall well-being.

In conclusion, frequency-based techniques have emerged as promising approaches for pain management and rehabilitation. By harnessing the power of specific vibrations and oscillations, these techniques can alleviate pain, promote healing, and enhance overall function. Whether through electrical stimulation, laser therapy, or sound waves, frequency-based techniques offer individuals a non-invasive and drug-free option for managing pain and facilitating rehabilitation.

11.3 Frequency-Based Approaches for Boosting Immune Function

The immune system plays a crucial role in protecting the body against harmful pathogens and maintaining overall health. It is responsible for identifying and eliminating foreign invaders, such as bacteria, viruses, and parasites, while also recognizing and neutralizing abnormal cells, including cancer cells. The immune system is a complex network of cells, tissues, and organs that work together to defend the body.

In recent years, there has been growing interest in the role of frequency in boosting immune function. Frequency-based approaches aim to stimulate and enhance the body's natural defense mechanisms, promoting a stronger immune response. By harnessing the power of frequency, individuals can potentially improve their immune function and reduce the risk of infections and diseases.

11.3.1 The Science Behind Frequency and Immune Function

To understand how frequency can boost immune function, it is important to delve into the science behind it. Every cell in the body has its own unique frequency or vibration. When the body is in a state of optimal health, these frequencies are balanced and harmonious. However, various factors such as stress, poor nutrition, and exposure to toxins can disrupt the natural frequencies, leading to imbalances and weakened immune function.

Frequency-based approaches for boosting immune function involve using specific frequencies to restore balance and harmony within the body. This can be achieved through various techniques, including sound therapy, electromagnetic therapy, and bioresonance therapy. These therapies aim to introduce specific frequencies into the body, either through external devices or through the body's own natural frequencies, to promote healing and enhance immune function.

11.3.2 The Impact of Frequency on Immune Function

Research suggests that frequency-based approaches can have a positive impact on immune function. Studies have shown that certain frequencies can stimulate the production of immune cells, such as lymphocytes and natural killer cells, which play a crucial role in fighting off infections and diseases. Additionally, frequency-based therapies have been found to enhance the activity of immune cells, improving their ability to recognize and eliminate pathogens.

Furthermore, frequency-based approaches can help regulate the body's inflammatory response. Inflammation is a natural immune response that helps the body fight off infections and repair damaged tissues. However, chronic inflammation can have detrimental effects on overall health and weaken the immune system. By using specific frequencies, it is possible to modulate the inflammatory response, promoting a balanced immune system and reducing the risk of chronic inflammation-related diseases.

11.3.3 Frequency-Based Techniques for Boosting Immune Function

There are several frequency-based techniques that can be used to boost immune function. These techniques aim to restore balance and harmony within the body, promoting optimal immune response. Some of these techniques include:

11.3.3.1 Sound Therapy

Sound therapy involves the use of specific frequencies and vibrations to promote healing and enhance immune function. This can be achieved through the use of sound healing instruments, such as singing bowls, tuning forks, or binaural beats. The vibrations produced by these instruments can help restore balance within the body, stimulating the immune system and promoting overall well-being.

11.3.3.2 Electromagnetic Therapy

Electromagnetic therapy utilizes electromagnetic fields to stimulate the body's natural healing processes and enhance immune function. This can be done through the use of devices that emit specific frequencies, such as pulsed electromagnetic field therapy (PEMF) devices. These devices deliver electromagnetic pulses to the body, promoting cellular regeneration and immune system activation.

11.3.3.3 Bioresonance Therapy

Bioresonance therapy involves the use of electromagnetic frequencies to detect and correct imbalances within the body. This therapy utilizes devices that measure the body's electromagnetic frequencies and then deliver specific frequencies to restore balance. By addressing underlying imbalances, bioresonance therapy can help boost immune function and promote overall health.

11.3.4 Frequency-Based Approaches for a Healthy Immune System

In addition to specific frequency-based techniques, there are several lifestyle factors that can support a healthy immune system. These include:

- Proper Nutrition: A balanced diet rich in vitamins, minerals, and antioxidants is essential for a strong immune system. Consuming a variety of fruits, vegetables, whole grains, lean proteins, and healthy fats can provide the necessary nutrients to support immune function.
- Regular Exercise: Engaging in regular physical activity can help boost immune function by improving circulation, reducing inflammation, and promoting the production of immune cells. Aim for at least 150 minutes of moderate-intensity exercise per week.
- Stress Management: Chronic stress can weaken the immune system. Incorporating stress management techniques, such as meditation, deep

breathing exercises, and yoga, can help reduce stress levels and support immune function.

- Sufficient Sleep: Sleep is crucial for immune function. Aim for 7-9 hours of quality sleep each night to allow the body to repair and regenerate. Establishing a regular sleep schedule and creating a relaxing bedtime routine can promote better sleep.
- Hygiene Practices: Practicing good hygiene, such as regular handwashing, can help prevent the spread of infections and support immune function.

By incorporating frequency-based approaches and adopting a healthy lifestyle, individuals can enhance their immune function and promote overall well-being. It is important to consult with healthcare professionals or trained practitioners when considering frequency-based therapies to ensure safe and effective use.

11.4 Frequency Techniques for Enhancing Overall Physical Health

Frequency plays a significant role in enhancing overall physical health. By understanding and harnessing the power of frequency, individuals can optimize their well-being and improve their physical health in various ways. In this section, we will explore different frequency techniques that can be used to enhance physical health and promote a healthier lifestyle.

11.4.1 Frequency and Physical Performance

One of the key areas where frequency can have a profound impact is on physical performance and endurance. Research has shown that certain frequencies can stimulate the body's energy systems, improve muscle coordination, and enhance overall athletic performance. By incorporating frequency-based techniques into training routines, individuals can experience increased stamina, improved strength, and enhanced physical performance.

11.4.2 Frequency for Pain Management and Rehabilitation

Frequency techniques can also be utilized for pain management and rehabilitation purposes. Certain frequencies have been found to have analgesic properties, helping to alleviate pain and discomfort. By targeting specific areas of the body with the appropriate frequencies, individuals can experience relief from chronic pain conditions, such as arthritis or fibromyalgia. Additionally, frequency-based therapies can aid in the rehabilitation process by promoting tissue healing and reducing inflammation.

11.4.3 Frequency-Based Approaches for Boosting Immune Function

Maintaining a strong immune system is crucial for overall physical health. Frequency-based approaches can be used to boost immune function and enhance the body's natural defense mechanisms. Research has shown that specific frequencies can stimulate the production of immune cells, improve lymphatic circulation, and enhance the body's ability to fight off infections. By incorporating frequency techniques into daily routines, individuals can support their immune system and reduce the risk of illness.

11.4.4 Frequency Techniques for Enhancing Cardiovascular Health

Cardiovascular health is essential for overall physical well-being. Frequency techniques can be utilized to improve cardiovascular function and promote heart health. Certain frequencies have been found to enhance blood circulation, reduce blood pressure, and improve the efficiency of the cardiovascular system. By incorporating frequency-based practices, such as meditation or breathing exercises, individuals can support their cardiovascular health and reduce the risk of heart-related conditions.

11.4.5 Frequency and Digestive Health

The digestive system plays a vital role in overall physical health. Frequency techniques can be used to promote digestive health and improve the efficiency of the digestive process. Certain frequencies have been found to stimulate the production of digestive enzymes, enhance nutrient absorption, and regulate bowel movements. By incorporating frequency-based practices, such as sound therapy or specific dietary frequencies, individuals can support their digestive health and improve overall well-being.

11.4.6 Frequency-Based Techniques for Detoxification

Detoxification is an essential process for eliminating toxins from the body and maintaining optimal physical health. Frequency-based techniques can aid in the detoxification process by stimulating the body's natural detoxification pathways. Certain frequencies have been found to enhance liver function, improve lymphatic drainage, and support the elimination of toxins from the body. By incorporating frequency-based practices, such as infrared saunas or specific frequency therapies, individuals can support their body's detoxification process and promote overall physical health.

11.4.7 Frequency and Bone Health

Maintaining strong and healthy bones is crucial for overall physical well-being, especially as individuals age. Frequency techniques can be used to support bone health and prevent conditions such as osteoporosis. Research has shown that certain frequencies can stimulate bone growth, enhance calcium absorption, and improve bone density. By incorporating frequency-based practices, such as low-intensity pulsed electromagnetic field therapy or specific frequency exercises, individuals can support their bone health and reduce the risk of bone-related conditions.

11.4.8 Frequency Techniques for Enhancing Flexibility and Mobility

Flexibility and mobility are essential for maintaining an active and healthy lifestyle. Frequency techniques can be utilized to enhance flexibility, improve joint mobility, and reduce the risk of injuries. Certain frequencies have been found to stimulate muscle relaxation, improve tissue elasticity, and promote joint flexibility. By incorporating frequency-based practices, such as stretching routines or specific frequency therapies, individuals can enhance their flexibility and mobility, leading to improved physical health.

Incorporating frequency techniques into daily life can have a profound impact on overall physical health. By utilizing the power of frequency, individuals can enhance their physical performance, manage pain, boost immune function, support cardiovascular health, improve digestive health, aid in detoxification, promote bone health, and enhance flexibility and mobility. These frequency-based practices provide individuals with the tools to optimize their physical well-being and lead a healthier, more vibrant life.

12 Incorporating Frequency into Daily Life

12.1 Practical Tips for Integrating Frequency Practices

Incorporating frequency practices into your daily life can have a profound impact on your overall well-being. By understanding the power of frequency and its effects on sleep, weight loss, and stress, you can optimize your lifestyle to promote better physical and mental health. In this section, we will explore some practical tips for integrating frequency practices into your daily routine.

12.1.1 Create a Frequency-Based Daily Routine

One of the most effective ways to integrate frequency practices into your life is by creating a daily routine that incorporates specific activities and techniques. Start by identifying the frequency-based practices that resonate with you the most, such as meditation, breathing exercises, or listening to frequency-enhanced music. Allocate dedicated time slots in your schedule for these practices, ensuring that you prioritize your well-being.

12.1.2 Start Small and Gradually Increase

When incorporating frequency practices into your daily routine, it's important to start small and gradually increase the duration and intensity of your practice. Begin with just a few minutes each day and gradually extend the time as you become more comfortable. This approach allows your mind and body to adapt to the new frequencies and ensures a sustainable integration of these practices into your life.

12.1.3 Find Frequency-Based Activities You Enjoy

To make frequency practices a regular part of your life, it's essential to find activities that you genuinely enjoy. Experiment with different techniques and practices to discover what resonates with you the most. Whether it's practicing yoga, engaging in mindful walking, or using frequency-enhanced essential oils, finding activities that bring you joy will make it easier to incorporate them into your daily routine.

12.1.4 Create a Sacred Space

Designating a specific area in your home as a sacred space for your frequency practices can enhance their effectiveness. This space should be free from distractions and clutter, allowing you to focus solely on your practice. Decorate it with items that inspire and uplift you, such as candles, crystals, or meaningful artwork. By creating a sacred space, you signal to your mind and body that it's time to engage in frequency practices and cultivate a sense of peace and tranquility.

12.1.5 Practice Mindfulness Throughout the Day

Integrating frequency practices into your daily life goes beyond dedicated practice sessions. It's important to cultivate mindfulness throughout the day, bringing your awareness to the present moment. Take moments to pause, breathe deeply, and tune in to the sensations in your body. This practice of mindfulness allows you to connect with the frequencies around you and promotes a sense of calm and centeredness.

12.1.6 Incorporate Frequency into Everyday Tasks

You can infuse frequency into your daily activities by incorporating it into everyday tasks. For example, while preparing meals, you can listen to frequency-enhanced music or affirmations to create a positive and harmonious environment. During your commute, you can engage in deep breathing exercises or listen to guided frequency meditations. By integrating frequency into these routine tasks, you maximize the benefits and make them an integral part of your daily life.

12.1.7 Seek Support and Accountability

Integrating frequency practices into your life can be a transformative journey, but it's essential to seek support and accountability along the way. Connect with like-minded individuals who are also interested in frequency practices, whether through online communities, local groups, or workshops. Share your experiences, learn from others, and hold each other accountable to maintain consistency in your practice.

12.1.8 Be Patient and Persistent

Integrating frequency practices into your daily life is a process that requires patience and persistence. It may take time for you to fully experience the benefits and see significant changes in your well-being. Be gentle with yourself and trust the process. Stay committed to your practice, even on days when it feels challenging or when you don't see immediate results. Consistency and perseverance will ultimately lead to long-term positive effects on your sleep, weight loss, and overall well-being.

Incorporating frequency practices into your daily routine is a powerful way to optimize your physical and mental health. By following these practical tips, you can seamlessly integrate frequency practices into your life and experience the transformative effects on your sleep, weight loss, stress management, and overall well-being. Embrace the power of frequency and embark on a journey of self-discovery and holistic healing.

12.2 Creating a Frequency-Based Daily Routine

Incorporating frequency practices into your daily routine can have a profound impact on your overall well-being. By intentionally engaging with specific frequencies, you can enhance various aspects of your life, including sleep, weight loss, stress management, mental health, and physical performance. Creating a frequency-based daily routine allows you to harness the power of these vibrations and optimize your daily activities for maximum benefits.

12.2.1 Understanding the Role of Frequency in Daily Life

Frequency is present in every aspect of our lives, from the natural rhythms of the earth to the electromagnetic waves that surround us. By understanding the role of frequency in daily life, you can begin to harness its power and incorporate it into your routine.

One way to do this is by aligning your activities with specific frequencies. For example, starting your day with a high-frequency activity such as meditation or yoga can help set a positive tone for the rest of the day. Similarly, ending your day with a low-frequency activity like reading or listening to calming music can promote relaxation and prepare your body for sleep.

12.2.2 Designing Your Frequency-Based Daily Routine

Designing a frequency-based daily routine involves intentionally incorporating activities that resonate with specific frequencies throughout your day. Here are some steps to help you create your own routine:

1. Identify your goals: Determine the areas of your life that you want to improve or enhance through frequency practices. Whether it's better

179

sleep, weight loss, stress reduction, or overall well-being, having clear goals will guide your routine design.

2. Research frequencies: Explore the frequencies that are known to have positive effects on your desired goals. For example, if you want to improve sleep, research frequencies that promote relaxation and deep sleep. If weight loss is your goal, look for frequencies that boost metabolism and fat burning.

3. Allocate time: Allocate specific time slots in your daily schedule for frequency-based activities. This could include morning and evening rituals, as well as incorporating frequency practices throughout the day. Be realistic about the time you can commit and ensure that it aligns with your lifestyle.

4. Morning routine: Start your day with high-frequency activities that energize and set a positive tone. This could include meditation, affirmations, visualization, or listening to uplifting music. Engaging with these activities can help you cultivate a positive mindset and enhance your overall well-being.

5. Throughout the day: Incorporate frequency practices into your daily activities. For example, you can listen to frequency-based music or guided meditations during your commute, take short breaks for deep breathing exercises, or practice mindfulness during meals. These small but intentional actions can help you stay connected to the desired frequencies throughout the day.

6. Evening routine: Wind down your day with low-frequency activities that promote relaxation and prepare your body for sleep. This could include practices such as journaling, gentle stretching, or listening to calming music. By engaging with these activities, you can signal to your body that it's time to unwind and promote a restful night's sleep.

7. Consistency is key: To experience the full benefits of a frequency-based daily routine, consistency is crucial. Make a commitment to stick to your routine and prioritize these activities in your daily life. Over time, the cumulative effects of consistent practice will become more apparent, leading to long-term improvements in your well-being.

12.2.3 Enhancing Mindfulness through Frequency

Incorporating frequency practices into your daily routine can also enhance your mindfulness and cultivate present moment awareness. Mindfulness involves paying attention to the present moment without judgment, and frequency can serve as an anchor to bring you back to the present.

By intentionally engaging with frequency-based activities, such as focusing on the vibrations of a singing bowl or tuning in to the rhythm of your breath, you can deepen your mindfulness practice. These activities help you become more attuned to the present moment and can enhance your overall well-being.

12.2.4 Sustaining Frequency Practices for Long-Term Well-Being

To sustain frequency practices for long-term well-being, it's important to integrate them into your daily routine in a way that feels natural and enjoyable. Here are some tips to help you maintain your frequency-based practices:

1. Start small: Begin by incorporating one or two frequency-based activities into your routine and gradually expand from there. Starting small allows you to build consistency and prevents overwhelm.
2. Find what resonates with you: Experiment with different frequency practices and find the ones that resonate with you the most. Whether it's listening to specific frequencies, engaging in energy healing practices, or using frequency-based devices, choose activities that you enjoy and that align with your goals.
3. Stay flexible: Be open to adapting your routine as needed. Life can be unpredictable, and it's important to be flexible and adjust your frequency practices to fit your changing circumstances.
4. Seek support: Connect with others who are also interested in frequency practices. Join online communities, attend workshops or

classes, or find a mentor who can guide you on your journey. Having support can help you stay motivated and inspired.

By creating a frequency-based daily routine and integrating it into your life, you can unlock the transformative power of frequency and experience profound improvements in your sleep, weight loss efforts, stress management, mental health, and overall well-being. Embrace the vibrations of frequency and let them guide you towards a healthier and more fulfilling life.

12.3 Frequency and Mindfulness

In our fast-paced and hectic lives, it is easy to get caught up in the constant stream of thoughts and distractions that pull us away from the present moment. However, by incorporating frequency and mindfulness practices into our daily routines, we can cultivate a greater sense of awareness and enhance our overall well-being.

12.3.1 Understanding Mindfulness

Mindfulness is the practice of intentionally bringing one's attention to the present moment without judgment. It involves being fully engaged in the here and now, rather than dwelling on the past or worrying about the future. By practicing mindfulness, we can develop a deeper understanding of ourselves and the world around us.

12.3.2 The Role of Frequency in Mindfulness

Frequency plays a crucial role in mindfulness practices as it helps to anchor our attention to the present moment. By focusing on the sensations and vibrations of the present moment, we can become more attuned to our thoughts, emotions, and physical sensations. This heightened awareness allows us to respond to life's challenges with greater clarity and compassion.

12.3.3 Cultivating Present Moment Awareness

One of the key benefits of incorporating frequency into mindfulness practices is the ability to cultivate present moment awareness. By tuning into the frequency of our breath, heartbeat, or the sounds around us, we can anchor our attention to the present moment. This helps to quiet the mind and reduce the tendency to get caught up in worries or regrets.

12.3.4 Frequency-Based Mindfulness Techniques

There are several frequency-based mindfulness techniques that can be incorporated into our daily lives to enhance present moment awareness. Here are a few examples:

12.3.4.1 Breath Awareness

Breath awareness is a fundamental mindfulness practice that involves focusing on the sensation of the breath as it enters and leaves the body. By tuning into the frequency of our breath, we can bring our attention back to the present moment whenever our mind starts to wander.

12.3.4.2 Body Scan

The body scan is a mindfulness practice that involves systematically bringing attention to different parts of the body. By tuning into the frequency of sensations in each body part, we can develop a deeper connection with our physical selves and cultivate a sense of groundedness in the present moment.

12.3.4.3 Sound Meditation

Sound meditation involves focusing on the sounds around us, whether it be the chirping of birds, the rustling of leaves, or the hum of traffic. By tuning into the frequency of these sounds, we can cultivate a sense of openness and curiosity, allowing us to fully immerse ourselves in the present moment.

12.3.5 The Benefits of Frequency and Mindfulness

Incorporating frequency and mindfulness practices into our daily lives can have a profound impact on our overall well-being. Here are some of the benefits:

12.3.5.1 Reduced Stress and Anxiety

By cultivating present moment awareness through frequency and mindfulness practices, we can reduce stress and anxiety. By focusing on the present moment, we can let go of worries about the future or regrets about the past, allowing us to experience a greater sense of calm and peace.

12.3.5.2 Improved Mental Clarity and Focus

Frequency and mindfulness practices help to quiet the mind and improve mental clarity and focus. By training our attention to stay in the present moment, we can enhance our ability to concentrate on tasks and make better decisions.

12.3.5.3 Enhanced Emotional Well-being

By cultivating present moment awareness, we can develop a greater understanding of our emotions and how they arise and pass away. This increased emotional intelligence allows us to respond to challenging situations with greater compassion and empathy, leading to enhanced emotional well-being.

12.3.5.4 Increased Self-Awareness

Frequency and mindfulness practices help us develop a deeper connection with ourselves. By tuning into the frequency of our thoughts, emotions, and physical sensations, we can gain a greater understanding of our inner world and develop a stronger sense of self-awareness.

12.3.6 Sustaining Frequency and Mindfulness Practices for Long-Term Well-Being

To sustain frequency and mindfulness practices for long-term well-being, it is important to integrate them into our daily routines. Here are some tips:

- Start with small steps: Begin by dedicating a few minutes each day to practice frequency and mindfulness. As you become more comfortable, gradually increase the duration of your practice.
- Create reminders: Set reminders or alarms throughout the day to bring your attention back to the present moment. This can help you develop a habit of mindfulness and maintain a frequency-based mindset.
- Find support: Join a mindfulness group or seek out like-minded individuals who can provide support and encouragement on your journey. Sharing experiences and insights can deepen your practice and keep you motivated.
- Be patient and kind to yourself: Remember that mindfulness is a lifelong practice, and it takes time to develop. Be patient with yourself and approach your practice with kindness and self-compassion.

By incorporating frequency and mindfulness practices into our daily lives, we can cultivate present moment awareness, reduce stress, enhance mental clarity, and improve overall well-being. Start small, be consistent, and enjoy the transformative effects of frequency and mindfulness in your life.

12.4 Sustaining Frequency Practices for Long-Term Well-Being

Incorporating frequency practices into your daily life can have profound effects on your overall well-being. However, to truly reap the benefits, it is important to sustain these practices for the long term. This section will explore strategies for maintaining frequency practices and maximizing their impact on your well-being.

12.4.1 Consistency is Key

Consistency is crucial when it comes to sustaining frequency practices. Just like any other habit, incorporating frequency into your daily routine requires dedication and commitment. Set aside specific times each day to engage in frequency-based activities, whether it's practicing mindfulness, using frequency techniques for sleep optimization, or incorporating frequency into your weight loss strategies. By making frequency a priority in your daily life, you can ensure that it becomes a sustainable practice.

12.4.2 Accountability and Support

Having a support system can greatly enhance your ability to sustain frequency practices. Share your goals and intentions with friends, family, or a community of like-minded individuals who are also interested in frequency-based practices. This support network can provide encouragement, motivation, and accountability, making it easier to stay on track with your frequency routines. Consider joining online forums, attending workshops, or even forming a frequency practice group to connect with others who share your interests.

12.4.3 Adaptability and Flexibility

While consistency is important, it is also essential to be adaptable and flexible in your frequency practices. Life can be unpredictable, and there may be times when it is challenging to maintain your usual routine. Instead of becoming discouraged, embrace the opportunity to explore new ways of incorporating frequency into your life. For example, if you are traveling and unable to engage in your usual frequency-based activities, try listening to frequency-enhanced music or guided meditations on your mobile device. By being open to different approaches, you can sustain your frequency practices even in the face of obstacles.

12.4.4 Tracking Progress and Celebrating Milestones

Tracking your progress and celebrating milestones can be a powerful motivator in sustaining frequency practices. Keep a journal or use a tracking app to record your frequency activities and note any changes or improvements you experience. This will not only help you stay accountable but also serve as a reminder of the positive impact frequency is having on your well-being. Celebrate your achievements along the way, whether it's reaching a specific weight loss goal, experiencing improved sleep quality, or managing stress more effectively. Recognizing and celebrating your progress will reinforce your commitment to sustaining frequency practices.

12.4.5 Continual Learning and Exploration

To sustain frequency practices for long-term well-being, it is important to continue learning and exploring new techniques and approaches. The field of frequency research is constantly evolving, and there are always new discoveries and advancements to explore. Stay curious and open-minded, seeking out new information, attending workshops, and reading books and articles on the subject. By expanding your knowledge and trying new frequency-based practices, you can deepen your understanding and enhance the benefits you derive from frequency.

12.4.6 Self-Care and Balance

Sustaining frequency practices for long-term well-being also requires a focus on self-care and balance. It is important to prioritize your physical, mental, and emotional health in order to maintain a sustainable frequency routine. Make time for activities that nourish your body and mind, such as exercise, healthy eating, relaxation techniques, and engaging in hobbies or activities that bring you joy. Remember that frequency practices are just one aspect of a holistic approach to well-being, and taking care of yourself in all areas of your life will support the sustainability of your frequency practices.

12.4.7 Reflection and Intention Setting

Regular reflection and intention setting can help you stay connected to your frequency practices and maintain a sense of purpose. Take time to reflect on the impact frequency has had on your well-being and set intentions for how you want to continue incorporating frequency into your life. This can be done through journaling, meditation, or simply taking a few moments each day to pause and connect with your intentions. By regularly reaffirming your commitment to sustaining frequency practices, you can stay aligned with your goals and maintain a sense of purpose in your frequency journey.

12.4.8 Seek Professional Guidance

If you find it challenging to sustain frequency practices on your own, consider seeking professional guidance. There are experts in the field of frequency who can provide personalized advice and support tailored to your specific needs and goals. Whether it's a frequency-based therapist, a sleep specialist, or a nutritionist with expertise in frequency-based weight loss strategies, working with a professional can provide valuable insights and help you navigate any challenges you may encounter along the way.

By implementing these strategies and approaches, you can sustain frequency practices for long-term well-being. Remember that frequency is a powerful tool that can positively impact your sleep, weight loss efforts, stress management, and overall well-being. Embrace the journey and enjoy the transformative effects of frequency in your life.

www.ingramcontent.com/pod-product-compliance
Lightning Source LLC
Chambersburg PA
CBHW050444290526
45786CB00006B/2156